Designers' handbook to Eurocode 1

Part 1: Basis of design

H. Gulvanessian and M. Holický

Thomas Telford, London

Published by Thomas Telford Publishing, Thomas Telford Services Ltd, 1 Heron Quay, London E14 4JD

First published 1996

Distributors for Thomas Telford books are
USA: American Society of Civil Engineers, Publications Sales Department, 345 East 47th Street, New York, NY 10017-2398
Japan: Maruzen Co. Ltd, Book Department, 3–10 Nihonbashi 2-chome, Chuo-ku, Tokyo 103
Australia: DA Books and Journals, 648 Whitehorse Road, Mitcham 3132, Victoria

A catalogue record for this book is available from the British Library

Classification
Availability: Unrestricted
Content: Guidance based on research and best current practice
Status: Refereed
User: Civil and structural engineering designers

ISBN: 0 7277 2524 6

Typeset by Techset Composition Limited, Salisbury, Wiltshire
Printed in Great Britain by Bookcraft (Bath) Ltd

Preface

Eurocode 1, Part 1, Basis of Design is the head document of the Eurocode suite and describes the principles and requirements for safety, serviceability and durability of structures and is intended to be used, for direct application, with the other parts of Eurocode 1 and the design Eurocodes 2 to 9. As such it is the key Eurocode document.

Aims and objectives of this Handbook

The principal aim of this Handbook is to provide the user with guidance on the interpretation and use of *Eurocode 1, Part 1, Basis of Design*. The Handbook also provides information on the status of ENV Eurocodes, their use with regard to National Application Documents and the progression of the ENV Eurocodes to EN status. In producing this Handbook the authors have endeavoured to provide explanations and commentary to the clauses in *Eurocode 1, Part 1, Basis of Design* for all the categories of users identified in the Foreword of the Eurocode. Although the design Eurocodes are primarily intended for the design of buildings and civil engineering works, Eurocode 1, Basis of Design is intended for the consideration of more categories of users, including:

- designers and contractors (as for the other Eurocodes) plus
- code drafting committees
- clients
- public authorities and other bodies who produce Regulations.

Layout of this Handbook

Eurocode 1, Basis of Design has a Foreword and nine Sections together with four informative Annexes. This Handbook has an Introduction which corresponds to the Foreword of Eurocode 1, Basis of Design and Chapters 1 to 9 which correspond to Sections 1 to 9. The numbering of Chapters and paragraphs in this Handbook corresponds to those in Eurocode 1, Basis of Design, for example: paragraph 9.4 of the Handbook is a commentary on clause 9.4 of the Eurocode. All cross-references in this Handbook to Sections, clauses, sub-clauses, annexes, figures and tables of Eurocode 1, Basis of Design are in italic type. The numbers for the expressions correspond unless prefixed by D (D for Designers' Handbook). Expressions prefixed by D do not appear in Eurocode 1, Basis of Design.

This Handbook has two types of appendix. Particular Chapters may have their own appendices which provide detailed background and further explanation to particular clauses; these appendices may be of interest only to a particular category of user. These are referred to, for example, as Appendix 1 to Chapter 5.

There are also four main appendices, A to D, which provide background and useful advice relating to the whole Handbook and should be of interest to all categories of users.

Acknowledgements

The authors are most grateful to the members of the Project Team, Eurocode 1, Part 1, Basis of Design, Dr. G. Breitschaft, Prof. H. Gulvanessian, Prof. N. Krebs Oversen, Mr. J. C. Leray, Prof. R. S. Narayanan, Prof. L. Ostlund, Prof. G. Sedlacek and Prof. A. Vrouwenvelder. In particular much reference was made to their Background Document[1] to Eurocode 1, Basis of Design. In addition, the authors thank Prof. R. S. Narayanan and Prof. A. W. Beeby who agreed that Examples 9.1. and 9.2 should be based on two of their examples from the Handbook[2] for Eurocode 2 in this series.

The authors also thank Mr. Roger Lovegrove and Dr. T. D. G. Cansius of the Building Research Establishment for their help and advice in drafting particular parts of this Handbook. They also wish to acknowledge the outstanding quality of the considerable secretarial support provided by Prof. Gulvanessian's personal assistant Mrs Carol Hadden, both for her work on Eurocode 1 and for this Handbook, and to thank the Building Research Establishment, Department of the Environment, UK, and the Klockner Institute of the Czech Technical University in Prague for the facilities provided.

October 1996

H. Gulvanessian
M. Holický

Contents

Introduction

The material in this chapter is covered in the Foreword to ENV 1991-1, *Eurocode 1, Basis of design and actions on structures, Part 1, Basis of design*, in clauses on

- Objectives of the Eurocodes
- Background to the Eurocode Programme
- The Eurocode Programme
- The technical objectives of *Eurocode 1, Basis of design*
- Intended users of *Eurocode 1, Basis of design*
- Intended uses of *Eurocode 1, Basis of design*
- Division into main text and annexes
- National Application Documents (NADs)
- Intended future developments of *Eurocode 1, Basis of design*.

The following abbreviations are used in this chapter

CEC	Commission of European Communities
CEN	European Committee for Standardisation
CPD	Construction Products Directive
EFTA	European Free Trade Association
EN	European Standard
ENV	European Pre-Standard
EU	European Union
PT	Project Team
SC	Sub-committee
TC	Technical Committee.

Objectives of the Eurocodes

Clauses 1 to 3

The Structural Eurocodes comprise a group of standards for the structural and geotechnical design of building and civil engineering works. They also cover execution and control to the extent that is necessary to indicate the quality of the construction products, and the standard of the workmanship, needed to comply with the assumptions of the design rules.

In addition, the Structural Eurocodes are intended to serve as reference documents for the following purposes:

(a) as a means to prove compliance of building and civil engineering works with the essential requirements of the Construction Products Directive (see Appendix 1)

(b) as a framework for drawing up harmonised technical specifications for construction products.

Background to the Eurocode programme

Clauses 4 to 6

Based on an agreement between CEC and CEN, the mandate to establish the Structural Eurocodes was given to CEN Technical Committee 250 (CEN/TC 250).

Eurocode programme

Clause 7

Eurocode 1, Part 1 lists the following Structural Eurocodes, each generally consisting of a number of parts which are in different stages of development at present.

EN 1991   *Eurocode 1: Basis of design and actions on structures*

EN 1992   *Eurocode 2: Design of concrete structures*

EN 1993   *Eurocode 3: Design of steel structures*

EN 1994   *Eurocode 4: Design of composite steel and concrete structures*

EN 1995   *Eurocode 5: Design of timber structures*

EN 1996   *Eurocode 6: Design of masonry structures*

EN 1997   *Eurocode 7: Geotechnical design*

EN 1998   *Eurocode 8: Design of structures for earthquake resistance*

EN 1999   *Eurocode 9: Design of aluminium alloy structures.*

The Structural Eurocodes are produced by separate Sub-committees (SCs) under the guidance and co-ordination of CEN/TC 250 (*clause 8*). The organisational structure of the Eurocode work is shown in Fig. 1.

Clause 8

Drafts for the Structural Eurocodes and their Parts are elaborated by Project Teams (PT in Fig. 1) which are selected by the appropriate SC. A Project Team consists of about six experts, who represent their SC. Delegates of the fifteen Member Organisations of the EU, the three EFTA countries and Associate EU member states are represented in CEN/TC 250 and its sub-committees. Voting is in accordance with the rules of CEN. The Member Organisations of the EU and the EFTA States have the right to vote.

Most of the Structural Eurocodes are at present in the ENV (European Pre-Standard) state (*clause 9*). With regard to *clauses 10 to 12*, CEN/TC 250 has recently issued the following Statement of Intent which gives information with regard to the transfer of the Pre-Standards to European Standards (EN).

Clause 9
Clauses 10 to 12

'The Technical Committee for the Structural Eurocodes, is determined to make rapid progress in producing and publishing the whole suite of Eurocodes, particularly the final EN documents which will be implemented by members in

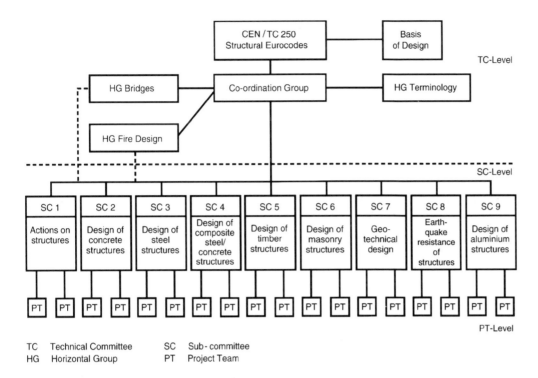

Fig. 1. Organisation of the Eurocode work

place of national standards for design and execution for building and civil engineering works.

The statement[3] that follows relates to those parts necessary for the design of buildings and gives a commitment to key dates for their achievement.

CEN/TC 250 is developing an action plan to achieve the following.

- All the EN Eurocodes necessary for the design of structures for buildings comprising Basis of Design, Loading and General Design Rules including those for Fire and Seismic Resistance, to be published by 1 January 2000.
- The first group of documents:

  EN 1991-1 *Eurocode 1, Basis of design and actions on structures, Part 1, Basis of design*

  EN 1991-2-1 *Eurocode 1, Basis of design and actions on structures, Part 2.1, Actions on structures, Densities, self-weight and imposed loads*

  EN 1991-2-3 *Eurocode 1, Basis of design and actions on structures, Part 2.3, Actions on structures, Snow loads*

  EN 1991-2-4 *Eurocode 1, Basis of design and actions on structures, Part 2.4, Actions on structures, Wind actions*

  EN 1992-1-1 *Eurocode 2, Design of concrete structures, Part 1.1, General rules, General rules and rules for buildings*

  EN 1993-1-1 *Eurocode 3, Design of steel structures, Part 1.1, General rules, General rules and rules for buildings*

EN 1994-1-1 *Eurocode 4, Design of composite steel and concrete structures, Part 1.1, General rules, General rules and rules for buildings* to be published before the end of 1998.

- The ENVs for Bridge Parts to be approved by the end of 1996 to allow their conversion to ENs before the end of 2001.'

It is possible that some SCs may not meet this programme and there may be some slippage. However, no slippage is expected with regard to the programme of Eurocode 1, Basis of design.

Technical objectives of Eurocode 1, Part 1, Basis of design

Eurocode 1, Part 1 describes the principles and requirements for safety, serviceability and durability of structures and is intended to be used for direct application with the other Parts of Eurocode 1 dealing with Actions on Structures and the design Eurocodes (Eurocodes 2 to 9). Fig. 2 shows the structure of the future European Standards System for building and civil engineering works, using *Eurocode 2, Design of concrete structures*, as an example.

Clauses 13 and 14

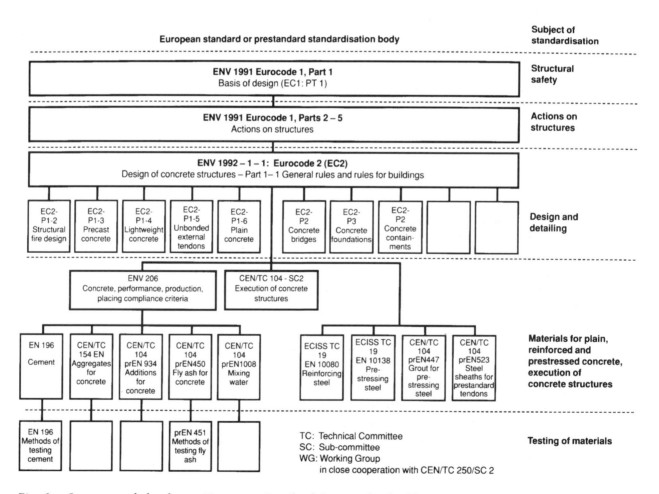

Fig. 2. Structure of the future European Standard System for building and civil engineering works using Eurocode 2 as an example

In addition, Eurocode 1, Part 1 provides guidelines for the aspects of structural reliability relating to safety, serviceability and durability for design cases not covered by the Eurocodes (e.g. other Actions, other materials and types of structures not treated), and to serve as a reference document for other CEN committees concerned with structural engineering aspects.

Clause 15

Intended users

Eurocode 1, Part 1 is intended for the consideration of more categories of users than are the other Eurocodes. It lists the intended users as:

- code drafting committees
- clients, for the specific requirements on reliability levels and durability
- designers
- contractors
- public authorities.

Each of these will have a different perspective on the provisions of Eurocode 1, Part 1.

Clause 17

Intended uses

Eurocode 1, Part 1 is intended for the design of structures within the scope of the Eurocodes (see Chapter 1). Additionally it can be used as a guidance document in the design of structures outside the scope of the Eurocodes for

Clauses 18 and 19

- assessing other actions and their combinations
- modelling material and structural behaviour
- assessing numerical values of the reliability format.

If Eurocode 1, Part 1 is used as a reference document by other CEN committees the numerical values provided in the document for safety factors should be used.

Clause 20

Division into main text and annexes

Eurocode 1, Part 1 is divided into a main text and a series of informative Annexes. The division takes into account the development expected during the transposition period from ENV to EN. The main text includes most of the principal and operational rules necessary for direct application of designs in the field covered by the Eurocodes.

Clauses 21 to 23

National Application Documents (NADs)

Clause 24

Clause 24 explains the relation of Structural Eurocodes and NADs. It is intended that, during the ENV period, Eurocode 1, Basis of design be used for design purposes, in conjunction with the particular NAD valid in the member state where the designed structures are to be located. Establishing the NAD is the responsibility of the National Competent Authority for the particular member state and, therefore, the purpose of the NAD is to provide essential information, for example on

- safety with due consideration of National Regulations
- the status and use of annexes

to enable Eurocode 1, Part 1 to be used for the design of buildings and civil engineering works in the particular member state. Hence, for example, a German designer using Eurocode 1 for the design of a building to be located in the UK must use the ENV, together with the UK NAD for Eurocode 1.

Appendix D of this Handbook gives the names and addresses of the appropriate National Standard Organisations to whom enquiries may be made with regard to the availability of NADs for a particular country.

Clause 25 particularly asks each country to confirm or amend the numerical values identified as 'boxed' or by [ ]. Some numerical values, for example partial safety factors and combination coefficients, are boxed in Eurocode 1, Part 1. These values are provided only as indications. Each member state is required to fix the boxed values applicable within its jurisdiction and state them in its NADs. In the long run it is hoped that many of these values can be harmonised across the European Community.   *Clause 25*

Each Structural Eurocode, or one of its parts, will normally have its unique NAD for each country.

Intended future development of Eurocode 1, Part 1, Basis of design

CEN/TC250, the head Committee overseeing the development of the Structural Eurocodes, has set up a Group to consider the transposition of Eurocode 1, Part 1 from ENV to EN status. The Group has recommended[4] that particular topics in Eurocode 1, Part 1 be developed as explained in Table 1. In addition to Table 1 the Group has recommended further developments on classification of structures.

Table 1. Recommended topics in Eurocode 1, Part 1 to be developed by transposition to EN 1991

Item	Current situation and need for improvement	First step (EN 1991-1: 1998)	Second step (EN 1991-1: 5 year revision)
Serviceability limit state	The requirements relating to serviceability criteria are well defined but the verification rules need improvement	The basic concepts for serviceability need broadening to define more accurately the purpose of the appropriate verifications with regard to the fundamental requirement	Development of material-independent performance criteria for the application parts of EN 1991-1
Static equilibrium	The guidance provided on static equilibrium is general and broader information on treating static equilibrium for all types of structures is required	Definition of static equilibrium, with broadening of the concept of static equilibrium in EN 1991-1 to take account of all types of structures during the execution process and normal use	Development of specific rules for all the application parts of EN 1991-1
Durability	The main requirement for defining durability is the design working life, and the guidance provided needs improvement	Table 2.1 to be developed in order to provide useful information for the design of structures and structural components	
Fatigue verification	ENV 1991-1 provides an Annex for fatigue which should be brought into the main text	Transfer Annex B of ENV 1991-1 to the main text with any appropriate rules from the design Eurocodes	Development of rules for the application parts of EN 1991-1
Structural analysis	The current guidance provided has to be broadened to be useful in design and to form a basis for harmonisation of the information in the design Eurocodes	The information in the design Eurocodes on structural analysis should be brought together and harmonised into ENV 1991-1 with advice on the application of these methods	
Annex A	Annex A is user-unfriendly and can lead to unsafe designs if used wrongly; it should be clarified, improved and completed	Completely re-edit and provide explanations for the differences between γ_m values in the design Eurocodes, and if possible harmonise with a common equation	
Soil structure/interaction	Application rules in Table 9.2, but ENV 1991-1 does not provide principles on this topic	The field of application for cases B and C of Table 9.2 should be defined more precisely	Production of comprehensive rules together with CEN/ TC250/SC7

CHAPTER 1

General

This chapter is concerned with the general aspects of *Eurocode 1, Part 1, Basis of design*. The material in this chapter is covered in *Section 1*, in the following clauses:

- scope   *clause 1.1*
- normative references   *clause 1.2*
- assumptions   *clause 1.3*
- distinction between Principles
  and Application Rules   *clause 1.4*
- definitions   *clause 1.5*
- symbols   *clause 1.6.*

1.1. Scope

Primary scope

Clause 1.1(1)

Eurocode 1, Part 1 is considered the head document in the Eurocode suite and it establishes for all the structural Eurocodes the principles and requirements for safety and serviceability of structures; it further describes the basis of design and verification and provides guidelines for related aspects of structural reliability.

Clause 1.1(2)

Most importantly, in addition to establishing the principles and requirements, it provides the basis and general principles for the structural design of buildings and civil engineering works (including geotechnical aspects) and must be used in conjunction with the Actions parts of Eurocode 1 and Eurocodes 2 to 9.

Scope in relation to design cases not covered by the Eurocodes

Clause 1.1(3)

As mentioned in the Introduction in the paragraphs on Intended uses, Eurocode 1, Part 1 provides guidelines for the aspects of structural reliability relating to safety, serviceability and durability for design cases not covered by the Eurocodes (e.g. other actions, other materials and types of structures not treated).

Scope in relation to structural design for execution stage and temporary structures

Clause 1.1(4)

Eurocode 1, Part 1 is also applicable to the structural design for the execution stage and to temporary structures. However, these two topics are not fully within the scope of Eurocode 1, Part 1, and therefore appropriate assumptions need to be made (e.g. amendments in partial factors for actions because of different design working life and different reliability levels).

Simplified methods of verification
Clause 1.1(5)

Simplified rules are provided in *Section 9*, but these are applicable to building structures only.

Scope in relation to bridges
Clause 1.1(6)

The majority of clauses in Eurocode 1, Part 1 are applicable for most types of civil engineering works including bridges. However, Eurocode 1, Part 1 does have building-specific clauses. The equivalents of these clauses for bridges are in Eurocode 1, Part 3.

Scope in relation to assessment of structures
Clause 1.1(7)

Clause 1.1(7) states that Eurocode 1, Part 1 is not directly aimed at the structural appraisal. There are generally three main methods of assessing existing structures, as follows.

(*a*) Check the structures against the design standards applicable at the time they were built. This method is generally possible only where either there is no change of use or the new use is demonstrably of similar or lower loadings.

(*b*) Assume material properties of the existing structure, based on experience, and then check design in accordance with current design codes. This method is applicable if the materials used are not covered by current codes but are broadly compatible in terms of strength, composition, ductility, etc. with modern materials.

(*c*) Test samples to derive equivalent 'modern' properties of the materials in situ and then check against current limit state methods and codes.

There are no current European codes for the appraisal of existing structures. In the UK, for example, the Building Regulations suggest that guidance can be obtained from the Institution of Structural Engineers publication *Appraisal of existing structures*[5] and BRE digest 366,[6] which discuss these approaches. Additionally, guidance is being developed in particular countries, e.g. Switzerland[7] and the Czech Republic.[8]

While Eurocode 1, Part 1 states that it can only offer guidance on the appraisal of existing structures, the approach given in Eurocode 1, Part 1 is broadly in line with method (*c*) (and to a lesser extent method (*b*)) above, and hence provides a crystallisation of good current UK practice, particularly since it combines the two methods. However, *clause 1.1(7)* notes that it is not directly intended for appraisal of existing structures, but may be so used where applicable.
Clause 1.1(7)

Scope in relation to special construction (e.g. nuclear structures)
Clause 1.1(8)

Clause 1.1(8) mentions that Eurocode 1, Part 1 does not completely cover the design of special structures which require unusual reliability considerations, citing as an example nuclear structures. Eurocode 1, Part 1 does not mention other structures that it may not cover completely, but offshore structures and particular structures concerned with defence works may also fall outside the scope.

Scope in relation to structures where deformations modify direct actions
Clause 1.1(9)

Eurocode 1, Part 1 does not cover the design of structures that are flexible to such an extent that deformations modify direct actions, and does not completely cover structures where interaction of actions and structural response is essential (e.g. second-order action effects and wind oscillations).

1.2. Normative references

No comment is necessary.

Clause 1.2

1.3. Assumptions

All the Structural Eurocodes, including Eurocode 1, Part 1 (*clause 1.3*), state the following assumptions associated with the validity of the design rules given in Eurocodes 1 to 9.

Clause 1.3

- The choice of the structural system and the design of a structure is made by appropriately qualified and experienced personnel.
- Execution is carried out by personnel having the appropriate skill and experience.
- Adequate supervision and quality control is provided during execution of the work, i.e. in design offices, factories, plants, and on site.
- The construction materials and products are used as specified in this Eurocode or in ENVs 1992 to 1999 or in the relevant supporting material or product specifications.
- The structure will be adequately maintained.
- The structure will be used in accordance with the design assumptions.
- Design procedures are valid only when the requirements for the materials, execution and workmanship given in ENVs 1992 to 1996 and 1999 are also complied with.

1.4. Distinction between Principles and Application Rules

The clauses in Eurocode 1, Part 1 are set out as Principles and Application Rules.

Clause 1.4(1)

Principles comprise general statements for which there is no alternative and requirements and analytical models for which no alternative is permitted unless specifically stated. Principles are distinguished by the prefix 'P' following the paragraph number. The verb 'shall' is always used in the Principle clauses.

Clause 1.4(2)

Clause 1.4(3)

Application Rules are generally acceptable methods, which follow the principles and satisfy their requirements. Alternative rules to those given in Eurocode 1, Part 1 are permissible provided that it can be demonstrated that they comply with the Principles and have at least the same reliability.

Clause 1.4(4)

Application rules have only a paragraph number. The verb 'should' is normally used for application rules. The verbs 'may' and 'can' have been used in isolated cases.

Clause 1.4(5)

1.5. Definitions

Most of the definitions given in Eurocode 1, Part 1 are reproduced from ISO 8930:1987.[9]

With regard to the definitions in *clause 1.5*, there are significant differences from usages in current National Codes and Standards (e.g. Definitions in British codes), to improve precision of meaning and to facilitate translation into other European languages.

Clause 1.5

In the Eurocode suite

- *action* means a load, or an imposed deformation (e.g. temperature effects or settlement)
- *strength* is a mechanical property of a material, in units of stress

- *resistance* is a mechanical property of a component or a cross-section of a member, or a member or structure
- *capacity* is a geometrical property, relevant to compatibility rather than to equilibrium (e.g. rotation capacity of a plastic hinge).

Action effects are internal moments and forces, bending moments, shear forces and deformations caused by actions.

The definitions given in Eurocode 1, Part 1 are subdivided in the following paragraphs:

- *clause 1.5.1* common terms used in the Structural Eurocodes
- *clause 1.5.2* special terms relating to design in general
- *clause 1.5.3* terms relating to actions
- *clause 1.5.4* terms relating to material properties
- *clause 1.5.5* terms relating to geometric data.

However, the definitions are difficult to locate even in the subparagraphs as they are not listed alphabetically. To help the reader, an alphabetical list referencing the clauses for a particular definition is provided in Appendix 1 to this chapter.

The following comments are made for particular definitions. Most comments take account of the definitions provided in ISO 2394, *General principles on reliability for structures*.[10] Where no alternative is provided, the definition in Eurocode 1, Part 1 accords with that in ISO 2394.

- *Clause 1.5.1.6.* The ISO definition provided in this clause is considered preferable and should be used. *Structure:* Organised combination of connected parts designed to provide some measure of rigidity and resistance against various actions.
- *Clause 1.5.2.4. Persistent design situations.* Note: This definition generally refers to conditions of normal use. Normal use includes possible extreme loading conditions from wind, snow, imposed loads, and earthquakes in areas of high seismicity.
- *Clause 1.5.2.6. Design working life.* Note: This definition also applies to structural elements. The 'assumed period' may be better expressed as the 'period assumed in design'.
- *Clause 1.5.2.7. Hazard.* Note: Gross human errors during design and execution are frequently occurring hazards.
- *Clause 1.5.2.10. Limit states.* Note: The definition from ISO 2394 'A specified set of states which separate desired states (no failure) from undesired states (failure)' is a more general definition and complements the definition provided in Eurocode 1, Part 1.
- *Clause 1.5.2.12. Serviceability limit state.* Note: The definition from ISO 2394 'A limit state concerning the criteria governing function related to normal use' is a very good complementary definition.
- *Clause 1.5.2.12.1.* The following definition being considered by the Working Group for the revision of ISO 8930 is considered preferable and correct and should be used. *Irreversible serviceability limit states.* Serviceability limit states, some consequences of exceeding which remain permanent when the actions that caused them are removed.
- *Clause 1.5.2.14. Maintenance.* Note: With regard to the definition, *working life* is the actual physical period of the structure during which it is used for an intended purpose with anticipated maintenance, while the *design working life* (*clause 1.5.2.6*) is an assumed period.

- *Clause 1.5.2.16.* The ISO definition provided in this clause is considered preferable and may be used. *Reliability:* The ability of a structure or structural element to fulfil the specified requirements, including the working life, for which it has been designed. Note: Reliability covers structural safety and serviceability, and can be expressed in terms of probability.
- *Clause 1.5.3.9. Single action.* Note: A more appropriate term would be *independent action.*
- *Clause 1.5.3.13.* The ISO definition provided in this clause is considered preferable and may be used. *Representative value of an action:* A value used for the verification of a limit state. Representative values consist of characteristic values, combination values, frequent values and quasi-permanent values.
- *Clause 1.5.3.14. Characteristic value of an action.* Note: ISO 2394[10] defines this term as follows, which can help the understanding of the Eurocode 1, Part 1 definition. 'A value chosen, in so far as it can be fixed on statistical bases, so that it can be considered to have a prescribed probability of being exceeded towards unfavourable values during a chosen reference period'.
- *Clause 1.5.3.17. Frequent value of a variable action.* Note: The definition being proposed by the Working Group for the revision of ISO 8930[9] is better expressed and states: 'Value determined, in so far as it can be fixed on statistical bases, so that: the total time, within a chosen period of time, during which it is exceeded for only a small given part of the chosen period of time; or the frequency of its exceedance is limited to a given value'.
- *Clause 1.5.3.18. Quasi-permanent value of a variable action.* Note: The part of the Eurocode 1, Part 1 definition '...during which it is exceeded is a considerable part of the chosen period of time' is given in the latest draft of ISO 2394 as '...during which it is exceeded is of the magnitude half the period'.
- *Clause 1.5.4.1.* The ISO 2394[10] definition provided in this clause is considered preferable and may be used. *Characteristic value of a material property:* An a priori specified fractile of the statistical distribution of the material property in the supply produced within the scope of the relevant material standard.

1.6. Symbols

The notation in *clause 1.6* is based on ISO 3898:1987.[11]

Clause 1.6

With regard to the notation for actions, it is stated above that actions refer not only to forces directly applied to the structure but also to imposed deformations. These are referred to as *indirect actions*, and the subscript 'ind' is used to identify them. Actions are further subdivided as permanent (G) (self-weight), variable (Q) (imposed loads, snow loads, etc.) and accidental (A).

Characteristic values of any parameter are distinguished by the subscript 'k'. Design values carry a subscript 'd'.

Appendix 1: Index of definitions

Clause 1.5

CHAPTER 2

Requirements

This chapter is concerned with fundamental requirements of *Eurocode 1, Part 1, Basis of Design*. The material in this chapter is covered in *Section 2*, in the following clauses:

- fundamental requirements   *clause 2.1*
- reliability differentiation   *clause 2.2*
- design situations   *clause 2.3*
- design working life   *clause 2.4*
- durability   *clause 2.5*
- quality assurance   *clause 2.6.*

2.1. Fundamental requirements

Principal requirements

There are three principal fundamental requirements concerning the bearing capacity for any structure and structural elements. These are covered by *clauses 2.1(1)P* and *2.1(3)P*, and may be summarised as follows.

Clause 2.1(1)P
Clause 2.1(3)P

The structure and structural elements should be designed, executed and maintained in such a way that during their intended life with appropriate degrees of reliability and in an economic way they will:

- perform adequately under all expected actions (serviceability limit state requirement)
- withstand extreme and/or frequently repeated actions occurring during their construction and anticipated use (ultimate limit state requirement)
- not be damaged by events such as fire, explosions, impact or consequences of human errors to an extent disproportionate to the original cause (robustness requirement).

The design should consider all the above requirements, since any may be decisive for appropriate structures or structural elements. These requirements may be generally interrelated and partly overlapping.

Serviceability and ultimate limit states requirements

The first two requirements, concerning serviceability and ultimate strength requirements in general, are mutually dependent. In many common cases, a structure that has sufficient strength also has sufficient stiffness. However, with

Clause 2.1(2)

the current trends of using advanced analytical techniques and higher strength materials, and with more emphasis on economy, this may not be true for particular structures and structural elements. For example, a large span structure may have sufficient strength but not have the specified stiffness. Thus, generally, due regard should be given to both safety and serviceability, including durability in both cases.

To clarify the possible relation between ultimate limit state and serviceability limit state requirements, Fig. 2.1. shows, in a simplified way, a typical load–deformation relationship of a given structural element. The maximum characteristic load $F_{k,max}$ due to the ultimate limit state requirement is derived from the ultimate design load using appropriate partial factors. $F_{ser,1}$ is the maximum admissible load derived for a given structural element from the serviceability requirement C_1 relating to deformation. $F_{ser,2}$ is the maximum admissible load derived from the serviceability requirement relating to deformation C_2. In many common cases the maximum characteristic load $F_{k,max}$ is less than the maximum admissible load due to serviceability requirement. In Fig. 2.1 this case corresponds to the limiting deformation C_2 and corresponding admissible load $F_{ser,2}$. If, however, the more severe deformation C_1 applies, the maximum admissible load $F_{ser,1}$ may be less than the maximum characteristic load $F_{k,max}$ derived from the ultimate load, therefore in that case the serviceability requirement is more severe than the ultimate limit state requirement.

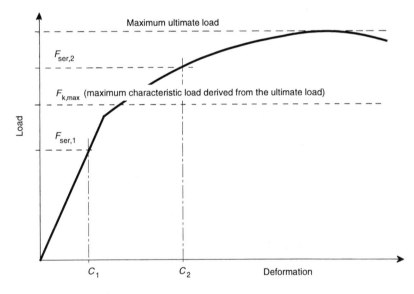

Fig. 2.1. Typical load–deformation relationship

Robustness requirements

The robustness requirement (*clause 2.1(3)P*) is additional to the serviceability and ultimate limit state requirements and refers to limiting the damage of a structure by events such as fire, explosion, impact or consequences of human error. To avoid damage or to ensure that damage is not disproportionate to the original

Clause 2.1(3)P

cause, Eurocode 1, Part 1 in *clauses 2.1(4)P* and *2.1(5)P* requires the appropriate choice of one or more of the following measures:

- avoiding, eliminating or reducing the hazards that the structure may sustain
- selecting a structural form that has low sensitivity to the hazards considered
- selecting structural form and design that survive adequately the accidental removal of an individual element or a limited part of the structure, or the occurrence of acceptable localised damage
- avoiding as far as is possible a structural system that may collapse without warning
- tying the structure together.

Little other practical guidance is provided in Eurocode 1 or Eurocodes 2, 3, 4, 5, 6 or 9 on ways of meeting these robustness requirements. There is, however, guidance based on UK practice, which is summarised below.[12] The guidance provided in the UK relates to avoiding disproportionate damage for

(1) buildings of more than four storeys and
(2) buildings that have a roof with a clear span exceeding 9 m between supports.

BUILDINGS OF MORE THAN FOUR STOREYS. To reduce the sensitivity of the building to disproportionate collapse in the event of an accident, the following approach is recommended.

(a) The provision of effective horizontal and vertical ties in accordance with the appropriate recommendations given in appropriate codes and Standards.[13-17] If these measures are followed then no further action is likely to be necessary according to UK guidance.
(b) If effective horizontal tying is provided and it is not feasible to provide effective vertical tying of any of the vertical load-bearing members, then each such untied member should be considered to be notionally removed, one at a time in each storey in turn, to check that its removal would allow the rest of the structure to bridge over the missing member, albeit in a substantially deformed condition. In considering this option, it should be recognised that certain areas of the structure (e.g. cantilevers or simply supported floor panels) will remain vulnerable to collapse. In these instances, the area of the structure at risk of collapse should be limited to that given in (c) below. If it is not possible to bridge over the missing member, that member should be designed as a protected member (see (d) below).
(c) If it is not feasible to provide effective horizontal and vertical tying of any of the load-bearing members, then the following accidental situation should be verified. Each support member should be considered to be notionally removed, one at a time in each storey in turn, and it should be checked that, on its removal, the area at risk of collapse of the structure within the storey and the immediately adjacent storeys is limited to 15% of the area of the storey or 70 m^2, whichever is the less (see Fig. 2.2). It should be noted that the area at risk is the area of the floor at risk of collapse on the removal of the member, and not necessarily the entire area supported by the member in conjunction with other members. If, on removal of a member, it is not possible to limit the area put at risk of collapse as above, that member should be designed as a protected member (see (d)).

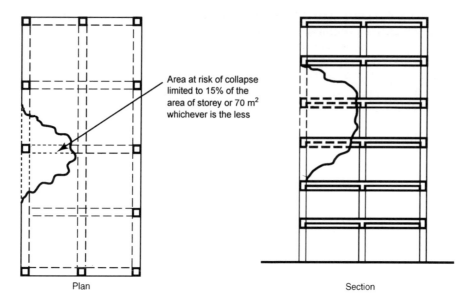

Fig. 2.2.   Area of risk of collapse in the event of an accident

(*d*)   The protected members (sometimes called 'key' elements) should be designed in accordance with the recommendations given in appropriate UK Codes and Standards.[13–17]

BUILDINGS THAT HAVE A ROOF WITH A CLEAR SPAN EXCEEDING 9 m BETWEEN SUPPORTS.   To reduce the sensitivity of the building to disproportionate collapse in the event of a local failure in the roof structure or its supports, the following approach is recommended.[12] Each member of the structure of the roof and its immediate supports should be considered to be notionally removed in turn one at a time, to check that its removal would not cause the building to collapse. In such circumstances it may be acceptable that:

(*a*)   other members supported by the notionally removed member collapse (see Fig. 2.3), and/or
(*b*)   the building deforms substantially.

Notwithstanding the above approach, consideration should be given to reducing the risks of local failure of the roof structure and its supports by:

(*a*)   protecting the structure from foreseeable physical damage
(*b*)   protecting the structure from adverse environmental conditions
(*c*)   making careful assessment and provision for movement and deformation of the structure
(*d*)   providing access for inspection of main structural components and joints.

2.2.   Reliability differentiation

Basic concepts

Eurocode 1, Part 1 explains in conceptual form the ways of achieving different 'levels of reliability' (*clause 2.2(1)P*), which has the same meaning as the term   *Clause 2.2(1)P* 'degrees of reliability' introduced in *clause 2.1(1)P*. This clause makes the very important statement that 'the reliability required for the majority of structures shall be obtained by design and execution according to Eurocodes, and appropriate quality assurance measures'. The term 'reliability' with regard to structural engineering should be considered as a structure's ability to fulfil its design purpose for some specified time.[10] In a narrow sense it is the probability that a structure will not exceed specified limit states (ultimate limit states and serviceability states) during a specified reference period (see Section 2.4).

Clause 2.1(1)P

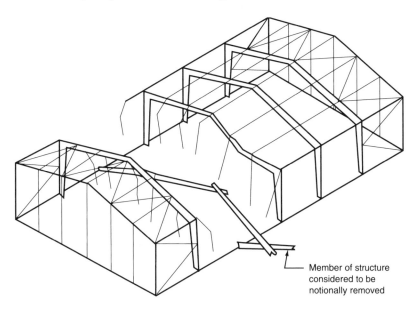

Member of structure considered to be notionally removed

Fig. 2.3. Acceptable extent of collapse in the event of a local failure in the roof structure a member supports

Choice of degree of reliability

Eurocode 1 allows the degree (level) of reliability to be adjusted (*clause 2.2(2)*) in the design, but the guidance provided for this is more conceptual than specific (*clause 2.2(3)*). The degree of reliability should be adopted so as to take account of:

Clause 2.2(2)

Clause 2.2(3)

- the cause and mode of failure—this implies that a structure or structural element that would be likely to collapse suddenly without warning (e.g. an element with low ductility) should be designed for a higher degree of reliability than one for which a collapse is preceded by some kind of warning in such a way that measures can be taken to limit the consequences
- the possible consequences of failure in terms of risk to life, injury, potential economic losses and level of social inconvenience
- the expense, level of effort and procedures necessary to reduce the risk of failure
- social and environmental conditions in a particular location.

Reliability differentiation and classification

The differentiation of the required degrees of reliability (*clause 2.2(4)*) in relation to structural safety and serviceability may be obtained by classification of whole structures or by classification of structural components. Thus, for example,

Clause 2.2(4)

degrees of reliability may be selected according to the consequences of failure as indicated by the following sequence of rules:[10]

- risk to life low, economic, social and environmental consequences small or negligible
- risk to life medium, economic, social or environmental consequences considerable
- risk to life high, economic, social or environmental consequences very great.

Table 2.1 indicates a possible classification (introducing an extremely high degree of reliability) of buildings and civil engineering works that may be used to select an appropriate degree of reliability according to consequences of failure.

Table 2.1. Examples of reliability differentiation according to the risk to life and risk of economic losses and social inconveniences

Degree of reliability	Risk to life, risk of economic and social losses	Examples of buildings and civil engineering works
Extremely high (use of clause 2.2(2))	Very High	Nuclear power reactors, major dams and barriers, srategic defence structures
Higher than normal (use of clause 2.2(2))	High	Significant bridges, grandstands, public buildings where consequences of failure are high
Normal (use of clause 2.2(1))	Medium	Residential and office buildings, public buildings where consequences of failure are medium
Lower than normal (use of clause 2.2(2))	Low	Agricultural buildings where people do not normally enter, greenhouses lightning poles

Recommended measures for reliability differentiation

Various possible measures by which the required reliability may be achieved include measures (*clauses 2.2(5) and (6)*) relating to design and quality assurance. There are attempts to check degrees of reliability across materials (e.g. structural steel and reinforced concrete) and different structures including geotechnical elements. At present, however, the reliability level is likely to be different in structures built in different materials.

Clauses 2.2(5) and 2.2(6)

2.3. Design situations

Variation of actions, environmental influences and structural properties, which occur throughout the life of the structure, should be considered in design by selecting distinct situations representing a certain time interval with associated hazards. These design situations should be selected so as to encompass all conditions that are reasonably foreseeable or occurring during the execution and use of the structure (*clauses 2.3(2)P and 2.3(3)*).

Four design situations are classified, as follows.

Clause 2.3(1)P

Clauses 2.3(2)P and 2.3(3)

(a) *Persistent situations* refer to conditions of normal use. These are generally related to the design working life of the structure. Normal use can include possible extreme loading conditions from wind, snow, imposed loads, etc.

(b) *Transient situations* refer to temporary conditions of the structure, in terms of its use or its exposure, e.g. during construction or repair. This implies the use of a time period much shorter than the design working life; one year may be adopted in most cases.

(c) *Accidental situations* refer to exceptional conditions of the structure or of its exposure, e.g. due to fire, explosion, impact, local failure. This implies the use of a relatively short period, but not for situations where a local failure may remain undetected.

(d) *Seismic situations* refer to exceptional conditions applicable to the structure when subjected to seismic events.

2.4. Design working life

The design working life is the term used for the assumed period for which a structure is to be used for its intended purpose *(clause 2.4(1)P)*. Required design working life is indicated in *Table 2.1* in *clause 2.4(2)*, and these values should be considered as indicative only. It is felt that, in its present form, *Table 2.1* needs improvement to be useful for practical guidance; an amended version, reproduced from the UK NAD[18] for Eurocode 1, Part 1, is given in Table 2.2. In comparison with ISO 2394[10] and some national standards, the periods indicated in *Table 2.1* seem to be rather low. Also, practical experience shows that the working life of many building structures, without major repair being necessary, is frequently longer.

Clause 2.4(1)P
Clause 2.4(2)

Table 2.2. Classification of design working life

Class	Design working life (years)	Examples
1	1–5	Temporary structures
2	25	Replacement structural parts, e.g. gantry girders, bearings
3	50	Buildings and other common structures other than those listed below
4	100	Monumental buildings and other special or important structures
5	120	Bridges

The present state of knowledge is insufficient to enable precise prediction of the life of a structure. The behaviour of materials and structures over extended periods of time can only be estimated. The likely period of maintenance of the structure or time of replacement of the various components of a structure can, however, be determined.

The notion of a design working life is useful for:

- the selection of design actions (e.g. imposed, wind, earthquake) and the consideration of material property deterioration (e.g. fatigue, creep)
- comparison of different design solutions and choice of materials, each of which will give a different balance between the initial cost and cost over an agreed period—life cycle costings will need to be undertaken to evaluate the relative economics of the different solutions

- evolving management procedures and strategies for systematic maintenance and renovation of structures.

Structures designed to the Eurocodes should perform and remain fit for the appropriate time, provided a maintenance (including replacement) strategy is developed by the client. In developing such a strategy the following aspects should be considered:

- costs of design, construction and use
- costs arising from hindrance of use
- risks and consequences of failure of the works during its working life and costs of insurance covering these risks
- planned partial renewal
- costs of inspections, maintenance, care and repair
- costs of operation and administration
- disposal
- environmental aspects.

2.5. Durability

The durability of a structure is its ability to remain fit for use during the design working life given appropriate maintenance (*clause 2.5(1)*). The structure should be designed in such a way that, or provided with protection so that, no significant deterioration is likely to occur within the period between successive inspections (*clause 2.5(2)*). The need for critical parts of the structure to be available for inspection, without complicated dismantling, should be considered in the design. Other interrelated factors that shall be considered to ensure an adequately durable structure are listed in *clause 2.5(3)P* and considered and explained below.

Clause 2.5(1)

Clause 2.5(2)

Clause 2.5(3)P

(a) *Intended and future use of the structure.* One example to consider is the abrasion on industrial floors due to machinery loads. The effects of a change of use on the durability of the structure should be considered, for example, when the micro-climate of a room (e.g. humidity in a laundry) or the exposure conditions may change.

(b) *Required performance criteria.* The design life requirement given in *clause 2.4* is the principal requirement to be considered in the overall strategy for achieving durability: in particular, decisions with regard to the life performance required from the structural elements and whether individual elements are to be replaceable, maintainable or should have a long-term design life.

Clause 2.4

(c) *Expected environmental influences.* The deterioration of concrete and timber and the corrosion of steel are affected by the environment and adequate measures need to be considered when considering the strategy to achieve durability. In addition, the variability of environmental actions, e.g. wind, snow and thermal actions, and their effects on the durability of a structure is an important consideration.

(d) *Composition, properties and performance of materials.* The use of materials that provide increased durability should be considered in the overall strategy for durability, with the use of preservative-treated timber, epoxy-coated reinforcing steels, stainless steel wall ties, concrete with low permeability, etc. For storage structures the choice of the material for the structure can be critical with regard to durability, for example for storing corrosive substances such as potash a glued laminated timber structural system is preferable to reinforced concrete or structural steel.

(e) *Choice of a structural system.* The structural form selected at the design stage should be robust, and the provision of redundancy in the structural system should be considered when designing for the consequences of known hazards. The design should avoid structural systems that are inherently vulnerable and sensitive to predictable damage and deterioration, and have flexibility 'built in' to enable the structure to tolerate changes in environmental conditions, movements, etc.

(f) *Shape of members and structural detailing.* The shape of members together with their detailing will influence the durability of a structure, for example an angle or channel steel section may retain or not retain moisture depending on its orientation.

(g) *Quality of workmanship and level of control.* The level of control of workmanship during execution can have an effect on the durability of a structure. For example, poor compaction can create honeycombs in reinforced concrete, and thus reduce durability.

(h) *Particular protective measures.* To increase durability, members should be protected from detrimental environments. For example, timber may be preservative-treated and/or given protective coatings, and steel members may be galvanised or clad with paints or concrete. Other measures such as cathodic protection of steel should also be considered.

(i) *Maintenance during the intended life.* Maintenance should be considered during the design and a strategy developed that is compatible with the design concept. Provision should be made for inspection, maintenance, and possible replacement if this is part of the performance profile. Whenever possible, a structure should be designed and detailed such that vulnerable, but important, members can be replaced without difficulty. An example is the provision for the replacement of post-tensioning tendons of pre-stressed concrete structures in corrosive environments.

The appropriate measures for various structural materials should be found in Eurocodes, i.e. 2 to 9 (*clause 2.5(4)*). Generally it is necessary (*clauses 2.5(5)P and (6)*) to appraise the environmental conditions and their significance in relation to durability at the design stage.

Clause 2.5(4)
Clauses 2.5(5)P and 2.5(6)

2.6.   Quality assurance

General

Eurocode 1, Part 1 in *clause 2.6.1* assumes that an appropriate quality policy is implemented by parties involved in the management of all stages in the life cycle of the construction process so as to fulfil the fundamental requirements described in Section 2.2. Practical experience shows that a quality system including organisation measures and control at the stages of design, execution, use and maintenance is the most significant tool to achieve an appropriate level of structural reliability.

Clause 2.6.1

The quality system of an organisation is influenced by the objectives of the organisation, by the product or services and by the practices specified by the organisation. Therefore, the quality system varies from one organisation to another. The CEN series EN 29 000 and the series of International Standards (ISO 9000 to 9004)[19-29] embody a rationalisation of the many and various national approaches in this sphere.

Specific aspects of quality policy in the field of construction works

A primary concern of a construction project is the quality of the construction works and, in particular, the reliability of the structure. The construction works should:

- meet a well-defined need, use or purpose
- satisfy client expectations
- comply with applicable standards and specifications
- comply with statutory (and other) requirements of society.

The objective of a quality policy is to meet these requirements.

Quality assurance through the construction works process life cycle

Quality assurance is an essential consideration in every stage in the design working life of any construction works. The various stages in a construction works life cycle, and the associated specific quality assurance activities, are shown schematically in the quality loop diagram in Fig. 2.4 and in Table 2.3.

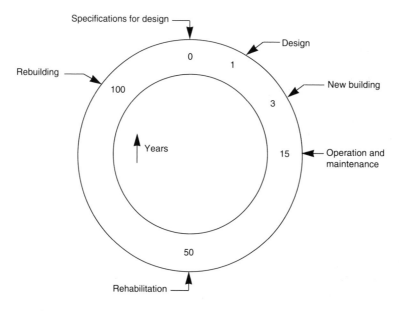

Fig. 2.4. Quality loop for buildings

Quality management

The quality management selected for implementing the quality policy should include consideration of:

- the type and use of the structure
- the consequences of quality deficiencies (e.g. structural failures)
- the management culture of the involved parties.

In the structural design of the construction works, reliability is the most important aspect to consider to achieve quality. Standards for structural design should provide a framework to achieve structural reliability as follows:

Table 2.3. Construction process and quality

Stage in quality loop	Activities
Conception	• Establishing appropriate levels of performance for construction works and components • Specification for design • Specification for suppliers • Preliminary specifications for execution and maintenance • Choice of intervening parties with appropriate qualifications for personnel and organisation
Design	• Specification of performance criteria for materials, components and assemblies • Confirmation of acceptability and achievability of performance • Specification of test options (prototype, in situ, etc.) • Specification for materials
Tendering	• Reviewing design documents, including performance specifications • Accepting requirements (contractor) • Accepting tender(s) (client)
Execution	• Reviewing process and product • Sampling and testing • Correction of deficiencies • Certification of work according to compliance tests specified in the design documentation
Completion of building and hand-over to client	• Commissioning • Verification of performance of completed building (e.g. by testing for anticipated operational loads)
Use and maintenance	• Monitoring performance • Inspection for deterioration or distress • Investigation of problems • Certification of work
Rehabilitation (or demolition)	• Similar to the above

- provide requirements for reliability
- specify the rules to verify the fulfilment of the requirements for reliability
- specify the rules for structural design and associated conditions.

The conditions to be fulfilled concern, for example, choice of structural system, the use of information technology in design through to execution, including supply chains with regard to materials being used, level of workmanship, and maintenance regime, and are normally detailed in the structural design standards. The conditions should also take into account the variability of material properties, the quality control and the criteria for material acceptance.

CHAPTER 3

Limit states

This chapter is concerned with the general concepts of limit states. The material in this chapter is covered in *Section 3*, in the following clauses:

- general   *clause 3.1*
- ultimate limit states   *clause 3.2*
- serviceability limit states   *clause 3.3*
- limit state design   *clause 3.4.*

3.1. General

Traditionally, according to the fundamental concept of limit states it is considered that the states of any structure may be classified as either satisfactory (safe, serviceable) or unsatisfactory (failed, unserviceable). Distinct conditions separating satisfactory and unsatisfactory states of a structure are called limit states. Thus, the limit states are those beyond which the structure no longer satisfies the design performance requirements (*clause 3.1(1)P*). Each limit state is therefore associated with a certain performance requirement. Often, however, these requirements are not formulated sufficiently clearly to allow for precise definition of appropriate limit states.

Clause 3.1(1)P

Generally, it may be difficult to express performance requirements qualitatively and to define the limit states unambiguously (particularly the ultimate limit states of structures made of ductile materials, and also the serviceability limit states, typically those affecting user comfort). In these cases, only a suitable approximation is available (e.g. the yield point of metals, or a limiting value for vertical deflection). These principles are indicated in Fig. 3.1 and provided here as background to the uncertainties of the limit state concept.[30–34] According to the traditional (sharp) concept of limit states described above, a given structure is assumed to be fully satisfactory up to a certain value of the load effect E_0 and beyond this value the structure is assumed to be fully unsatisfactory (see Fig. 3.1(a)). However, it may be very difficult to define precisely such a distinct value E_0, separating desired and undesired structural conditions, and the simplification in Fig. 3.1(a) may not be adequate. In these cases a transition region $< E_1, E_2 >$ in which a structure is gradually losing its ability to perform satisfactorily provides a more realistic (vague) concept (see Fig. 3.1(b)). Uncertainties in the vague concept of the limit states may be taken into account only in reliability analyses using special mathematical techniques that are not covered in the present generation of Eurocodes.

In order to simplify the design procedure, two fundamentally different types of

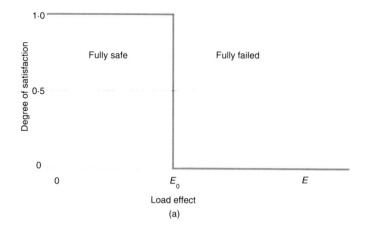

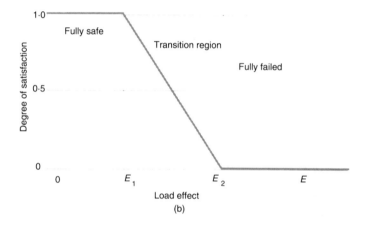

Fig. 3.1. *Sharp and vague definitions of a limit state*

limit states are generally recognised (*clause 3.1(2)*):

Clause 3.1(2)

(a) ultimate limit states
(b) serviceability limit states.

Ultimate limit states are associated with collapse or other similar forms of failure. Serviceability limit states correspond to conditions of normal use (deflections, vibration, cracks, etc.). In general the design should include both safety and serviceability, including durability in both cases (*clause 2.1(2)*). However, as indicated in Section 2.1 (see also Fig. 2.1), verification of one of the two limit states may be omitted if sufficient information is available to ensure that the requirements of one limit state are met by the other limit state.

Clause 2.1(2)

The nature of ultimate limit states is essentially different from the nature of serviceability limit states. There are two main reasons for this distinction:

(a) while the infringement of the ultimate limit states almost always leads to the overall loss of structural reliability and to the removal or fundamental repair of the structure, infringement of the serviceability limit states does not usually lead to such fatal consequences for the structure, and the structure may normally be used after the removal of the actions that caused the infringement

(b) while criteria of ultimate limit states involve parameters of a structure and appropriate actions only, criteria of serviceability limit states depend additionally on requirements of the client and users (sometimes very subjective), and characteristics of installed equipment or non-structural elements.

Differences between the ultimate limit states and serviceability limit states result in separate formulation of reliability conditions, and dissimilar reliability levels are assumed in verification of the two types of limit states.

As indicated in Section 2.3 (*clause 2.3(1)P*), variation of actions, environmental influences and structural properties, which occurs throughout the life of the structure, should be considered in design by selecting distinct situations (persistent, transient, accidental and seismic) representing a certain time interval with associated hazards (*clause 3.1(3)*). The ultimate and serviceability limit states should be considered in all these design situations (*clause 2.3(3)*), which should be selected to encompass all conditions that can reasonably be foreseen to occur during the execution and use of the structure. If two or more independent loads act simultaneously, their combination should be considered in accordance with Chapter 9 (see also *Section 9*). Within each load case, a number of realistic arrangements should be assumed to establish the envelope of action effects, which should be considered in the design.

Clause 2.3(1)P

Clause 3.1(3)
Clause 2.3(3)

3.2. Ultimate limit states

The ultimate limit states are associated with collapse and other similar forms of structural failure (*clause 3.2(1)P*). In almost all cases that concern the ultimate limit states the first passage of the limit state is equivalent to failure. In some cases, e.g. when excessive deformations are decisive, states prior to structural collapse can, for simplicity, be considered in place of the collapse itself and treated as ultimate limit states (*clause 3.2(2)*). As indicated in Section 3.1, ultimate limit states concern the safety of the structure and its contents and the safety of people (*clause 3.2(3)P*). These important circumstances should be taken into account when specifying reliability parameters of structural design and quality assurance.

Clause 3.2(1)P

Clause 3.2.(2)

Clause 3.2(3)P

The list of ultimate limit states provided in *clause 3.2(4)*, which may require consideration in the design, may be extended as follows:

Clause 3.2(4)

(a) loss of equilibrium of the structure or any part of it, considered as a rigid body
(b) failure of the structure or part of it due to rupture, fatigue or excessive deformation
(c) instability of the structure or one of its parts
(d) transformation of the structure or part of it into a mechanism
(e) sudden change of the structural system to a new system (e.g. snap-through).

Time-dependent structural properties such as fatigue and other time-dependent deterioration mechanisms reduce the strength of a structure and can initiate one of the above-mentioned ultimate limit states. In this respect it is useful to distinguish two types of structures: damage-tolerant (i.e. robust) and damage-intolerant (sensitive to minor disturbance or construction imperfections). Effects of various deteriorating mechanisms on the ultimate limit states should then be taken into account according to the type of the structure. Adequate reliability of damage-intolerant structures can also be assured by an appropriate quality control programme. In the case of damage-tolerant structures, fatigue damage may be regarded as a serviceability limit state.

3.3. Serviceability limit states

The serviceability limit states are associated with conditions of normal use (*clause 3.3(1)P*). They are concerned with the performance of the construction works or part of it, with the comfort of people and with the appearance of the construction works (*clause 3.3(2)P*). *Clause 3.3(1)P*

Clause 3.3(2)P

Taking into account the time-dependency of load effects, it is useful to distinguish two types of serviceability limit states which are illustrated in Fig. 3.2 (*clause 3.3(3)P*): *Clause 3.3(3)P*

(a)   irreversible serviceability limit states (see Fig. 3.2(a)), i.e. limit states that remain permanently exceeded even when the actions that caused the infringement are removed (e.g. a permanent local damage, permanent unacceptable deformations)

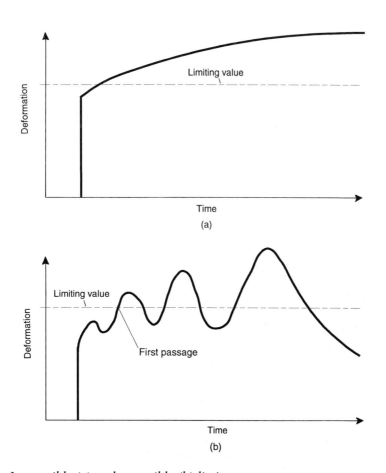

Fig. 3.2.   Irreversible (a) and reversible (b) limit states

(b) reversible serviceability limit states (see Fig. 3.2(b)), i.e. limit states that will not be exceeded when the actions that caused the infringement are removed (e.g. cracks of prestressed components, temporary deflections, excessive vibration).

For irreversible limit states the design criteria are similar to those of ultimate limit states. The first passage of the limit state is decisive (see Fig. 3.2). This important aspect of the irreversible limit states should be taken into account when determining serviceability requirements in contract or design documenta-tion (*clause 3.3(4)*).

Clause 3.3(4)

For reversible limit states the first infringement does not necessarily lead to failure and the loss of serviceability. Various serviceability requirements can be formulated taking into account the acceptance of infringements, their frequency and their duration. Generally, three types of serviceability limit states are applicable, as follows:

(a) no infringement is accepted
(b) specified duration and frequency of infringements are accepted
(c) specified long-term infringement is accepted.

The correct serviceability criteria are then associated as appropriate with the characteristic, frequent and quasi-permanent values of variable actions (see Section 4.3). The following combinations of actions corresponding to the above three types of limit states are generally used in verification of serviceability limit states for different design situations (see Chapter 9):

(a) the characteristic (rare) combination if no infringement is accepted
(b) the frequent combination if the specified time period and frequency of infringements are accepted
(c) the quasi-permanent combination if the specified long-term infringement is accepted.

The serviceability limit states (*clause 3.3(5)*) affecting the appearance, or effective use of, the structure which may require consideration in the design can be summarised as follows:

Clause 3.3(5)

(a) excessive deformation, displacement, sag and inclination which can affect, for example, the appearance of the structure
(b) excessive vibration (acceleration, amplitude, frequency) which can, for example, cause discomfort to people
(c) local damage and cracking which can affect durability of the structure
(d) observable damage due to fatigue and other time-dependent effects.

3.4. Limit state design

The design procedure using the limit state concept consists of setting up structural and load models for relevant ultimate and serviceability limit states which are considered in the various design situations and load cases (*clause 3.4(1)P*).

Clause 3.4(1)P

In most cases the design values should be determined by using the character-istic or representative values as defined in *Chapters 4–6* and *Sections 4–6* in combination with partial factors, and in combination with other coefficients (*clause 3.4(2)*).

Clause 3.4(2)

It should be emphasised that the direct determination of design values (*clause 3.4(3)*) should be used exceptionally when well-defined models and sufficient data are available (see also Chapter 8 and *Section 8*). In this case, as stated in

Clause 3.4(3)

clause 3.4(3), the design values should be chosen cautiously and should correspond to at least the same degree of reliability for various limit states as implied in the partial factors methods in Eurocode 1, Part 1.

Appendix 1: Limiting deflection for serviceability limit states

Informative limiting values for vertical deflection, horizontal deflection, and settlement of foundations are given as a guide in Tables 3.1, 3.2 and 3.3 respectively. The limiting values given in Tables 3.1 and 3.2 may be considered in order to verify serviceability limit states using the appropriate *expressions (16), (17) and (18)* of *clause 9.5*.

Clause 9.5

Table 3.1 gives informative limiting values for vertical deflections δ_{max} and δ_2 (defined below) of the horizontal member shown in Fig. 3.3, where δ_0 is the pre-camber (hogging) of the beam in the unloaded state (state (3)), δ_1 is the deflection of the beam due to the permanent loads immediately after loading (state (1)), δ_2 is the deflection of the beam due to the variable loads plus any time-dependent deformations due to the permanent load (state (3)), δ_{max} is the maximum sagging in the final state relative to the straight line joining the supports given as

$$\delta_{max} = \delta_1 + \delta_2 - \delta_0 \qquad \text{(D3.1)}$$

and L is the length of the span of beams supported at both ends and twice the length of cantilever projections.

Table 3.1. Recommended limiting values for vertical deflections

Condition		δ_{max}	δ_2
(a)	Roofs generally	$L/200$	$L/300$
(b)	Roofs with access other than for maintenance	$L/250$	$L/300$
(c)	Floors generally	$L/250$	$L/300$
(d)	Floors/roofs supporting plaster or other brittle finish or non-flexible partition	$L/250$	$L/350$
(e)	Floors supporting columns (unless the deflection has been taken into account within global analysis for the ultimate limit state)	$L/400$	$L/500$
(f)	When δ_{max} can impair the appearance of the building	$L/250$	—

See Fig. 3.3 for definition of δ_{max}, δ_2 and L.

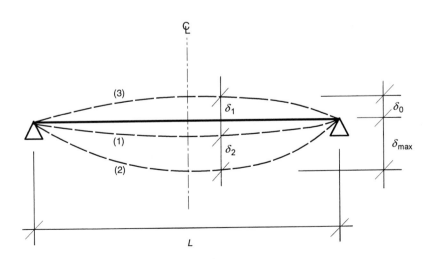

Fig. 3.3. Vertical deflections of a horizontal beam

Table 3.2. Recommended limiting values for horizontal deflection

Condition	δ
(a) Portal frames without gantry girders or masonry infills	$h/150$
(b) Other single-storey buildings	$h/300$
(c) Multi-storey buildings	
Each storey	$h/300$
Over the whole height of building	$h_0/500$

h is the height of the column or storey, h_0 is the overall height of the building.

Table 3.3. Recommended limiting values for settlement of foundation (see Fig. 3.4)*

Condition	δ
(a) Total settlement	
Isolated foundation	25 mm
Raft foundation	50 mm
(b) Relative settlement between adjacent columns	
Open frames	20 mm
Frames with flexible cladding or finishes	10 mm
Frames with rigid cladding or finishes	5 mm
(c) Relative rotation β	1/500
(d) Tilt ω	†

* Quasi-permanent values of variable and accidental actions should be considered when calculating settlement. Limiting values apply to foundation on sand. Higher limiting values may be permitted for foundation on clay soil provided the limits for β and ω are respected (see Fig. 3.4).
† To be determined by the designer.

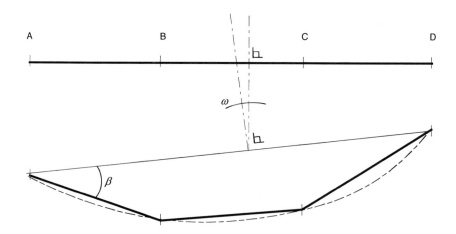

Fig. 3.4. Relative rotation β and tilt ω of foundation

Appendix 2: Vibration considerations for serviceability limit states

The vibration criteria may address three categories of 'receivers'[35] as follows:

(a) human occupants — including adjacent property
(b) building contents — including adjacent property
(c) building structure — including adjacent property.

Vibration criteria for human reaction to vibration can be further classified as pertaining to:

(a) sensitive occupancies such as hospital operating rooms
(b) regular occupancies such as offices and residential areas
(c) active occupancies such as assembly areas or places of heavy industrial work.

Vibration criteria due to human occupancy are given in terms of acceptance criteria.[36–38] These criteria include the relevant acceleration–frequency line for selected exposure time and direction of vibration.

The vibration criteria selected for the contents of buildings should assure satisfactory functioning of sensitive instruments or certain manufacturing processes. Because of the large variety of such equipment and processes, it is not possible to present any fixed levels of vibration amplitudes that ensure satisfactory operation. Limits for the movements of the machines are usually specified by considering the maximum deflection and frequency.

The vibration criteria selected for building structures should avoid the development of minor damage of structural and non-structural elements. The permissible levels of vibration effects depend on the type of structure, age, importance and other aspects.[36–38] Corresponding limits not covered by acceleration–frequency lines or deflection–frequency lines may be expressed in terms of the maximum stress, the maximum stress range or maximum deformation. These limits should be specified in the design specifications.

CHAPTER 4

Actions

This chapter is concerned with the actions and environmental influences on structures and civil engineering works. The material in this chapter is covered in *Section 4*, in the following clauses:

- principal classification   *clause 4.1*
- characteristic values of action   *clause 4.2*
- other representative values of variable and accidental actions   *clause 4.3*
- environmental influences   *clause 4.4.*

4.1. Principal classification

Classification of actions introduced in Eurocode 1, Part 1 (*clauses 4.1(1)P and 4.1(2)P*) provide the basis for modelling actions and controlling structural reliability. The aim of the classification is to identify the dissimilar character of various actions and to enable the use of appropriate theoretical action models and reliability elements in structural design. A complete action model describes several properties of the action, such as its magnitude, position, direction and duration. In some cases interactions between actions and the response of the structure should be taken into account (e.g. wind oscillations, soil pressures and imposed deformations).

Clauses 4.1(1)P and 4.1(2)P

The classification introduced in Eurocode 1, Part 1 takes into account the following aspects of actions and environmental influences:

(a) application of actions
(b) variation of actions in time
(c) variation of actions in space
(d) nature of actions and structural response.

Considering the nature of their application, all actions are allocated to two fundamental classes (*clause 4.1(1)P*) as direct actions (e.g. forces, loads) and indirect actions (e.g. imposed or constrained deformations due to temperature changes, moisture variation and uneven settlement or imposed acceleration due to machine excitation and earthquake). While models for direct actions can usually be determined independently of the structure properties, the representation of indirect actions can be determined only by taking structural response into account. The dissimilar character of the two classes of action leads to distinct theoretical models with different parameters (including the mean, standard deviation and other statistical parameters).

Clause 4.1(1)P

Considering their variation in time, all actions are classified as (*clause 4.1(2)P*):

Clause 4.1(2)P

(*a*) permanent actions G, where the variation in time is small and gradual, e.g. self-weight, weight of fixed equipment and surfaces

(*b*) variable actions Q, which consist of sustained action and intermittent actions, e.g. imposed loads, wind loads or snow loads

(*c*) accidental actions A, which occur extremely rarely and for a short period of time only, e.g. fire, impact loads (see Table 4.1).

Table 4.1. Classification of actions

Permanent action		Variable action		Accidental action	
(*a*)	Self-weight of structures, fittings and fixed equipment	(*a*)	Imposed floor loads	(*a*)	Explosions
(*b*)	Prestressing force	(*b*)	Snow loads	(*b*)	Fire
(*c*)	Water and soil pressures	(*c*)	Wind loads	(*c*)	Impact from vehicles
(*d*)	Indirect action, e.g. settlement of supports	(*d*)	Indirect action, e.g. temperature effects		

In design calculations different partial factors and other coefficients are applied for permanent, variable and accidental actions (see Chapter 9 and *Section 4.9*).

Considering their variation in space all actions are classified as:

(*a*) fixed actions, e.g. self-weight

(*b*) free actions, e.g. imposed loads, wind loads, snow loads.

In design calculations the fixed actions cannot be relocated, and the free action should be located in the most unfavourable position in accordance with the influence area to obtain an extreme load effect.

Considering the nature of actions and the structural response, all actions are classified as:

(*a*) static actions, which do not cause significant acceleration of the structure or structural member

(*b*) dynamic actions, which cause significant acceleration of the structure or structural member.

Often, dynamic effects of actions are considered as quasi-static actions by increasing the magnitude of the static actions or by the introduction of an equivalent static action (*clause 4.1(3)*). Some variable actions, static or dynamic, may cause stress fluctuation which may lead to fatigue of structural materials.

Clause 4.1(3)

Assignment of a specific action to the above-mentioned classes may depend on a particular design situation. Some actions, for example seismic actions and snow loads, can be considered as accidental and/or variable actions (*clause 4.1(4)*) depending on the site location. For example, *Eurocode 1, Part 3, Snow loads* allows the treatment of snow loads as an accidental action for some cases. This applies for specific climatic regions where the local drifting of snow on roofs is considered to form exceptional snow loads because of the rarity with which they occur, and these loads are treated as accidental loads in accordance with Eurocode 1, Part 1.

Clause 4.1(4)

Prestressing force P is normally considered as a permanent action (*clause 4.1(5)*); more detailed information on this is provided in Eurocodes 2, 3 and 4. Indirect actions are either permanent, e.g. settlement of supports, or variable, e.g. temperature effects (*clause 4.1(6)*, see also Table 4.1), and should be treated accordingly. The design values of these actions and their application in combinations should follow rules for the appropriate class of actions as defined in *Section 4.9* (see also Chapter 9). *Clause 4.1(5)*

Clause 4.1(6)

As mentioned above, the complete action model describes several properties of the action, such as its magnitude, position, direction and duration. In most cases the magnitude of an action is described by one quantity. For some actions a more complex representation of magnitudes may be necessary, e.g. for multi-dimensional actions, dynamic actions and variable actions causing fatigue of structural materials (*clause 4.1(7)*). For example, in the case of fatigue analysis it is necessary to identify a complete history of stress fluctuations, often in statistical terms, or to describe a set of stress cycles and the corresponding number of cycles. *Clause 4.1(7)*

In addition to the above classification of actions, Eurocode 1, Part 1 considers the environmental influences of a chemical, physical and biological character as a separate group of actions (*clause 4.4*). These influences have many aspects in common with the mechanical actions; in particular, they may be classified according to their variation in time as permanent (e.g. chemical impacts), variable (e.g. temperature and humidity influences) and accidental (e.g. spread of aggressive chemicals). Generally, the environmental influences may cause time-dependent deterioration of material properties and therefore may lead to gradually decreasing reliability of structures. *Clause 4.4*

4.2. Characteristic values of actions

All actions including environmental influences are introduced in design calculations by various representative values. The main representative value of an action F (*clause 4.2(1)P*) is the characteristic value F_k. Depending on available data and experience the characteristic value should be specified in the relevant Eurocodes as a mean, an upper or lower value or a nominal value (which does not refer to any statistical distribution). Exceptionally, the characteristic value of an action can also be specified in the design or by the relevant competent authority, provided that the general provisions specified in Eurocode 1 are observed (*clause 4.2(2)P*). *Clause 4.2(1)P*

Clause 4.2(2)P

There is a lack of relevant statistical data concerning various actions and environmental influences. Consequently, the determination of a representative value of an action may involve not only evaluation and analysis of available observations and experimental data, but often, when there is a complete lack of sufficient statistical data, a fairly subjective assessment, judgement (e.g. for particular accidental actions) or decision (e.g. for loads permitted on existing structures). For these cases, when a statistical distribution is not known, the characteristic value will be specified as a nominal value (*clause 4.2(7)P*). Nevertheless, all the characteristic values, regardless of the methods used for their original determination, are treated in the same way in Eurocode 1, Part 1. *Clause 4.2(7)P*

Determination of permanent actions

With regard to determining permanent actions G, particularly for the determination of self-weight for traditional structural materials, sufficient statistical data may be available. If the variability of a permanent action is small, a single characteristic value G_k may be used. When the variability of a permanent action is not small, two values have to be used: an upper value $G_{k,sup}$ and a lower value $G_{k,inf}$ (*clause 4.2(3)P*). The variability of the permanent action can usually be assumed to be small if the coefficient of variation during the design life is not greater than 0·1 (recommended value). However, if the structure is very sensitive to variations in G (e.g. some types of prestressed concrete structures) two values have to be considered even if the coefficient of variation is small (*clause 4.2(4)*). For the case when a single value may be used, G_k can be assumed to be the mean value μ_G (see Fig. 4.1; definitions of basic statistical parameters are given in Appendix B). In other cases, when two values are to be used, a lower value $G_{k,inf}$ and an upper value $G_{k,sup}$, representing the 0·05 and 0·95 fractiles respectively, as shown in Fig. 4.1, should be used. The normal (Gaussian) distribution (*clause 4.2(5)*) may be generally assumed for self-weight.

In the latter case the following relationships[39] (see also Appendix C for definitions of statistical terms and basic statistical techniques involved) can be used to determine the lower value $G_{k,inf}$ and the upper value $G_{k,sup}$

$$G_{k,inf} = \mu_G - 1{\cdot}64\sigma_G = \mu_G(1 - 1{\cdot}64V_G) \qquad (D4.1)$$

$$G_{k,sup} = \mu_G + 1{\cdot}64\sigma_G = \mu_G(1 + 1{\cdot}64V_G) \qquad (D4.2)$$

where V_G is the coefficient of variation of G. It follows from the above equations (see also Fig. 4.1) that for a coefficient of variation of 0·10 (which is a hypothetical boundary between a low and a high variability of permanent actions) the $G_{k,inf}$ and $G_{k,sup}$ will therefore be 16·4% less or greater than the mean value μ_G.

Clause 4.2(3)P

Clause 4.2(4)

Clause 4.2(5)

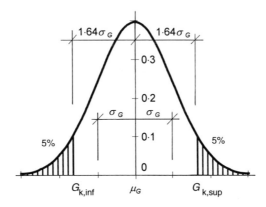

Fig. 4.1. *The lower value $G_{k,inf}$ or the upper value $G_{k,sup}$*

The self-weight of the structure can usually be represented by a single characteristic value and calculated on the basis of the nominal dimensions and mean unit mass (*clause 4.2(6)*), using values provided in *Eurocode 1, Part 2.1, Densities self-weight and imposed loads*. There may be special cases for particular design situations (e.g. when considering overturning and strength of retaining walls) when both the lower value $G_{k,inf}$ and the upper value $G_{k,sup}$ should be used in design (see also Chapter 9).

Clause 4.2(6)

Determination of variable actions

Considerably fewer statistical data are available for variable actions Q than for permanent actions. Consequently the characteristic value Q_k may be determined on the basis of a statistical model, if this is available, or by a specified nominal value if it is not (*clause 4.2(7)P*).

Clause 4.2(7)P

In the first case the characteristic value Q_k corresponds either to an upper value with an intended probability of it not being exceeded (the most common case), or to a lower value with an intended probability of it not falling below, during an assumed reference period. Hence, two separate elements are used to define the characteristic value: the reference period during which the extreme (maximum or minimum) is observed, and the intended probability with which this extreme value should not exceed or should not fall below the characteristic value. The characteristic values for individual types of variable actions (e.g. imposed load, snow load, wind load, traffic load) are provided in Eurocode 1, Parts 2.1, 2.3, 2.4 and 3.

For the determination of the characteristic value Q_k it is recommended by Eurocode 1, Part 1 that one consider the intended probability of it not being exceeded as 0·98, and a reference period as 1 year (*clause 4.2(8)*). These values correspond approximately to a return period (expected period between two subsequent occurrences) of the characteristic value being exceeded of 50 years. According to Eurocode 1, Part 1, depending on the character of variable action a different reference period may be more appropriate (e.g. 100 years for the case of actions on bridges), and this will change the intended probability of the characteristic value not being exceeded.

Clause 4.2(8)

Variable actions caused by water should be calculated by allowing for fluctuating water levels and for variation in appropriate geometric parameters, such as the profile of the structure or of components exposed to the water (*clause 4.2(9)*).

Clause 4.2(9)

Determination of accidental actions

Less statistical information is available for accidental actions than for permanent and variable actions. The appropriate clauses (*clauses 4.2(10)–(14)*) may be very simply summarised as follows. The characteristic values A_k are specified for fire actions in *Eurocode 1, Part 2.2, Action on structures exposed to fire*; and for explosions and some impacts in *Eurocode 1, Part 2.7, Accidental actions* (*clauses 4.1(10), 4.2(11) and 4.2(12)*); values are given for seismic actions in *Eurocode 8, Earthquake-resistant design of structures* (*clause 4.2(13)*), and for accidental actions (*clause 4.2(14)*) on bridges due to traffic in *Eurocode 1, Part 3, Traffic load* on bridges.

Clauses 4.2(10)–(14)

Clauses 4.1(10), 4.2(11) and 4.2(12)

Clause 4.2(13)

Clause 4.2(14)

For multi-component actions and some design situations the characteristic action is represented by groups of values, to be considered alternatively in design calculations (*clause 4.2(15)*, see also *clause 4.1(7)P*).

Clause 4.2(15)

Clause 4.1(7)P

4.3.   Other representative values of variable and accidental actions

In addition to the characteristic values of actions, other representative values are specified for variable and accidental actions in Eurocode 1, Part 1. Three representative values are commonly used for variable actions: the combination value $\psi_0 Q_k$, the frequent value $\psi_1 Q_k$ and the quasi-permanent value $\psi_2 Q_k$ (*clause 4.3(1)P*). The factors ψ_0, ψ_1 and ψ_2 reduce the characteristic values of

Clause 4.3(1)P

variable actions in order to take account of a reduced probability of simultaneous occurrence of the most unfavourable values of several independent actions.

For the persistent and transient design situations of ultimate limit states and for the characteristic (rare) combinations of serviceability limit states, only the non-dominant variable actions may be reduced using the ψ coefficients. In other cases (for accidental design situation and combinations of serviceability limit states), the dominant as well as non-dominant actions may be reduced using the ψ coefficients (see Table 4.2 and Chapter 9). If it proves difficult to decide which action is dominant (when considering combination of actions), a comparative study will be needed.

Table 4.2. Applications of coefficients ψ_0, ψ_1 and ψ_2 for dominant and non-dominant variable actions at ultimate and serviceability limit states*

Limit state	Design situation or combination	ψ_0	ψ_1	ψ_2
Ultimate	Persistent and transient	Non-dominant	X	X
	Accidental	X	Dominant	X
	Seismic	X	X	Non-dominant
Serviceability	Characteristic	Non-dominant	X	X
	Frequent	X	Dominant	Non-dominant
	Quasi-permanent	X	X	Dominant and non-dominant

* X, not applied.

Recommended values for all the three coefficients ψ_0, ψ_1 and ψ_2 are given in *clause 9.4.4* (see also Chapter 9). Their application in the verification of ultimate limit states and serviceability limit states is indicated in Table 4.2.

The combination value $\psi_0 Q_k$, the frequent value $\psi_1 Q_k$, and the quasi-permanent value $\psi_2 Q_k$ are shown schematically in Fig. 4.2. The combination value $\psi_0 Q_k$ is associated with the combination of actions for ultimate and irreversible serviceability limit states in order to take account of the reduced probability of simultaneous occurrence of the most unfavourable values of several independent actions (*clause 4.3(2)P*). A statistical technique to determine ψ_0 is presented in *Annex A*. The coefficient ψ_0 for combination values of some imposed and traffic actions in building is typically equal to 0·7 (see Chapter 9); this value is indicated approximately in Fig. 4.2.

The frequent value $\psi_1 Q_k$ is primarily associated with the frequent combination in the serviceability limit states. The frequent value $\psi_1 Q_k$ is also assumed to be appropriate for the verification of the accidental design situation of the ultimate limit states. In both cases the reduction factor ψ_1 is applied as a multiplier of the dominant variable action. The frequent value $\psi_1 Q_k$ of a variable action Q is determined so that the total time, within a chosen period of time, during which $Q > \psi_1 Q_k$ is only a specified (small) part of the period, or the frequency of the event $Q > \psi_1 Q_k$ is limited to a given value (*clause 4.3(3)P*). The total time for which $\psi_1 Q_k$ is exceeded is equal to the sum of time periods Δt_1, Δt_2, ... shown in Fig. 4.2 by thick parts of the horizontal line indicating the frequent value $\psi_1 Q_k$.

Eurocode 1, Part 1 recommends that for ordinary buildings the specified part of a chosen period be 0·05, or the frequency be 300 occurrences per year (*clause 4.3(4)*). These recommended values may be altered depending on the type of

Clause 9.4.4

Clause 4.3(2)P

Clause 4.3(3)P

Clause 4.3(4)

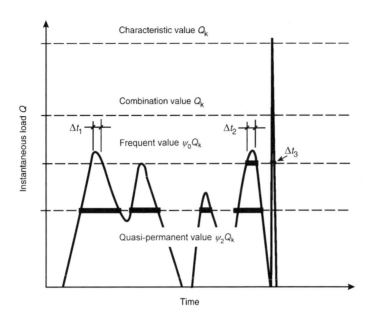

Fig. 4.2. Representative values of variable actions

construction works and design situations considered in the design. In some cases two (upper and lower) or more frequent values may be used. The coefficient ψ_1 for the frequent value of some imposed and traffic actions in building is equal to 0·5 (see Chapter 9); this value is indicated approximately in Fig. 4.2.

The quasi-permanent values $\psi_2 Q_k$ are used for the verification of accidental and seismic situations at the ultimate limit states and for verification of frequent and quasi-permanent combinations (long-term effects) of serviceability limit states.

The quasi-permanent values $\psi_2 Q_k$ were initially used in quasi-permanent combinations at serviceability limit states for all variable actions. Accordingly, the quasi-permanent value $\psi_2 Q_k$ is defined such that the total time within a chosen period of time during which it is exceeded, i.e. when $Q > \psi_2 Q_k$, is a considerable part (0·5) of the chosen period of time (*clauses 4.3(5)P and 4.3(6)*). The value may also be determined as the value averaged over the chosen period of time.

The same quasi-permanent values $\psi_2 Q_k$ are assumed to be appropriate also for accidental and seismic design situations of ultimate limit state and for the frequent combination of serviceability limit states (see Table 4.1). The total time of $\psi_2 Q_k$ being exceeded is equal to the sum of periods, shown in Fig. 4.2 by thick parts of the horizontal line indicating the quasi-permanent value $\psi_2 Q_k$. In some cases the coefficient ψ_2 for the frequent value of imposed and traffic actions is equal to 0·3 (see Chapter 9); this value is indicated approximately in Fig. 4.2.

The representative values $\psi_0 Q_k$, $\psi_1 Q_k$ and $\psi_2 Q_k$ and the characteristic values are used to define the design values of the actions and the combinations of actions as explained in Chapter 9 (*clause 4.3(7)P*). Their application in verification of the ultimate limit states and serviceability limit states is indicated in Table 4.2. Representative values other than those described above may be required for specific structures and special types of load, e.g. for fatigue load (*clause 4.3(8)*).

Clauses 4.3(5)P and 4.3(6)

Clause 4.3(7)P

Clause 4.3(8)

4.4.   Environmental influences

Environmental influences (carbon dioxides, chlorides, humidity, fire) may considerably affect the material properties and consequently safety and serviceability of structures in an unfavourable way. These effects are strongly material-dependent and therefore their characteristics have to be specified individually for each material in Eurocodes 2 to 9 (*clause 4.4*). When the environmental influences can be described by theoretical models and numerical values, the degradation of the material can be estimated by calculation.

Clause 4.4

In many cases, however, this may be difficult and only an approximate assessment can be made. Often combinations of various environmental influences may be decisive for the design of a particular structure in a given condition. More information on how to apply the representative values is available in Chapter 9 and in *Section 9*. Other detailed rules concerning the use of representative values for various structures are given in Eurocodes 2 to 9 (see also *clause 4.3(7)P*).

Clause 4.3(7)P

CHAPTER 5

Material properties

This chapter is concerned with various properties of construction materials used in the design of building structures and civil engineering works. The material in this chapter is covered in *clauses 5(1)P, 5(2), 5(3)P, 5(4), 5(5) and 5(6)* in *Section 5*. Information on these principles and application rules is presented in Section 5.1 below. The appendices provide general information about modelling of material properties (Appendix 1) and basic statistical techniques for specification of the characteristic and design values of material properties (Appendix 2).

Clauses 5(1)P, 5(2), 5(3)P, 5(4), 5(5) and 5(6)

5.1. Material properties in design calculation

Properties of materials and soils are an important group of basic variables for determining structural reliability. In design calculations the properties of materials (including soil and rock) or products are represented by characteristic values which correspond to the prescribed probability not being infringed (*clause 5(1)P*). Usually the 5% fractile is considered for strength parameters, while stiffness parameters are defined as the mean value (*clause 5(2)*). General information on the strength and stiffness parameters is given in Appendix 1 to this chapter.

Clause 5(1)P

Clause 5(2)

The operational rules to determine specified fractiles as described in Appendix 2 to this chapter can be used if the theoretical model for random behaviour of a material property is known or sufficient data are available to determine such a model. If, however, only limited test data are available, statistical uncertainty should be taken into account and the above-mentioned operational rules should be replaced by more complicated statistical techniques (see also Chapter 8).

A material property will normally be determined from standardised tests performed under specified conditions (*clause 5(3)P*). It is often necessary to apply a conversion factor to convert the test results into values that can be assumed to represent the behaviour of the structure or the ground. These factors and other details are given in Eurocodes 2 to 9. For traditional materials, e.g. steel and concrete, previous experience and extensive tests are available and appropriate conversion factors are well established and presented in various design codes. Properties of new materials should be obtained from an extensive testing programme, including tests on complete structures, revealing relevant properties and appropriate conversion factors. New materials should be introduced only if comprehensive information about their properties, supported by experimental evidence is available.

Clause 5(3)P

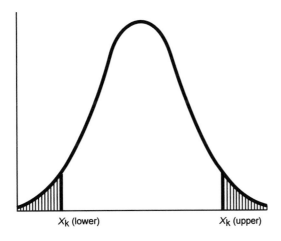

Fig. 5.1.   Lower and upper characteristic values

A material strength may have two characteristic values—an upper and a lower. For most cases the lower value (see Fig. 5.1) will need to be considered in the design.

There are, however, cases when an upper estimate of strength is required (e.g. for the tensile strength of concrete for the calculation of the effect of indirect actions). In these cases the use of a nominal upper value of the strength should be considered (*clause 5(4)*).

Clause 5(4)

When the lower or upper characteristic value is derived from tests the available data should be examined for cases where material is classified using a grading system (e.g. timber and steel components) with a number of classes, to avoid the possibility that a manufacturer may have included failed specimens for the upper grade into the lower grade, thus distorting the statistical characteristics (including the mean, standard deviation and fractiles) of the lower grade (see Fig. 5.2).

According to *clause 5(5)*, whenever there is a lack of information about the statistical distribution of the property a nominal value may be used in the design. In the case of insignificant sensitivity to the variability of the property a mean

Clause 5(5)

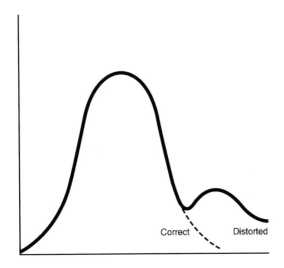

Fig. 5.2.   Distorted distribution

value may be considered as the characteristic value. Relevant values of material properties and their definitions are available in Eurocodes 2 to 9 (*clause 5(6)*).   *Clause 5(6)*

Appendix 1: Modelling of material properties

Generally the material properties should be described by measurable physical quantities corresponding to the properties considered in the calculation models. As a rule these physical quantities are time-dependent and may additionally depend on temperature, humidity, load history and environmental influences. They also depend on specified conditions concerning manufacturing, supply and acceptance criteria. The material properties and their variation should be determined from tests using standardised test specimens. The tests should be based on random samples that are representative of the population under consideration. By means of appropriately specified conversion factors or functions, the properties obtained from test specimens should be converted to properties corresponding to the assumptions made in calculation models. The uncertainties of the conversion factors should also be considered in calculation models. Conversion factors should cover size-effects, shape-effects, time-effects and the effects of temperature, humidity and other influences.

For soils, as for existing structures, the materials are not produced but are found in situ. Therefore, the values of the properties have to be determined for each project using appropriate tests. A detailed investigation based on test results may then provide more precise and complete information than would purely statistical data, especially with respect to systematic trends or weak spots in the spatial distributions. However, fluctuations in homogeneous materials and the limited precision of the tests and of their physical interpretations can be treated by statistical methods. For these materials the extent of investigation is an important element of the structural reliability, which is often difficult to quantify.

At the design stage, assumptions need to be made concerning the intended materials for the construction works. Therefore, corresponding statistical parameters have to be deduced from previous experience and existing populations that are considered appropriate to the construction works. The chosen parameters should be checked at the execution stage for quality using suitable (preferably statistical) methods of quality control. The identification of sufficiently homogeneous populations (e.g. appropriate divisions of the production in batches) and the size of samples are other important elements of the structural reliability.

The most important material properties introduced in design calculation describe the fundamental engineering aspects of building materials:

(*a*) strength f
(*b*) modulus of elasticity E
(*c*) yield stress (if applicable) σ_y
(*d*) limit of proportionality ε_y
(*e*) strain at rupture ε_f.

Figure 5.3 shows a typical example of a one-dimensional stress σ–strain ε diagram together with the above-mentioned fundamental quantities.

In design calculations, strength parameters are usually introduced by the lower 5% fractiles representing the characteristic values; the stiffness parameters are introduced by their mean values of these parameters (see Section 5.1). However, when stiffness affects the structural load-bearing capacity, the appropriate choice of the relevant partial factor should compensate for the mean value choice. As mentioned in Section 5.1, in some cases the upper characteristic value for the strength is of importance (e.g. for the tensile strength of concrete when the effect of indirect action is calculated). Some of the above-mentioned properties, e.g. strain at rupture, may be introduced in design calculations implicitly by appropriate conditions for validity of theoretical models of cross-section or structural member behaviour.

In addition to the basic material properties indicated in the one-dimensional stress σ–strain ε diagram (see Fig. 5.3), other important aspects need to be considered:

(*a*) multi-axial stress condition (e.g. Poisson's ratio, yield surface, flow and hardening rules, crack creation and crack behaviour)

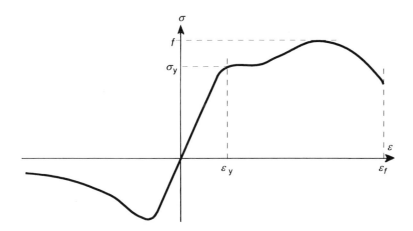

Fig. 5.3.   One-dimensional σ–ε diagram

(b)   temperature effects (e.g. coefficient of expansion, effect on materials properties including extreme conditions)
(c)   time effect (e.g. effect of internal and external influences, creep, creep rupture, consolidation of soils, fatigue deterioration)
(d)   dynamic effects (e.g. mass density and material damping, effect of loading rate)
(e)   humidity effects (e.g. shrinkage, effect on strength, stiffness and ductility)
(f)   effects of notches and flaws (e.g. unstable crack growth (brittle fracture), stress intensity factor, effect of ductility and crack geometry, toughness).

A very significant material property is the fatigue behaviour of structural members. This time-dependent effect of repeated loading on structural members is generally investigated by simplified tests, where the members are subjected to load variations of constant amplitude until excessive deformations or fracture due to cracks occur. The fatigue strength is then defined by characteristic $\Delta\sigma$–N curves that represent the 5% fractiles of failure. The test evaluation is carried out in accordance with *Annex D* and appropriate provisions of Eurocodes 2 to 9.

The characteristic $\Delta\sigma$–N curves are normally represented in a double logarithmic scale as indicated in Fig. 5.4. The corresponding equation has the form:

$$\Delta\sigma^{m}N = C = \text{constant} \tag{D5.1}$$

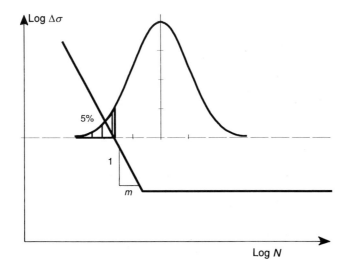

Fig. 5.4.   The characteristic $\Delta\sigma$–N curve

where $\Delta\sigma$ is the stress range calculated from the load range using appropriate material and geometrical properties, taking account of stress concentration factors, and N is the number of cycles. In some cases the stress concentration factors are introduced as an explicit coefficient of the nominal stress range.

Generally, the $\Delta\sigma$–N curves are dependent on geometric stress concentration and metallurgical aspects. Further details concerning the characteristic $\Delta\sigma$–N curves for different materials may be found in Eurocodes 2 to 9.

Appendix 2: Statistical technique for determination of the characteristic value

As most material properties are random variables of considerable scatter, applied values should always be based on appropriate statistical parameters or fractiles. Commonly, for a given material property x the following statistical parameters are considered: the mean μ_x, standard deviation σ_x, coefficient of skewness (asymmetry) α_x or other statistical parameters, e.g. lower or upper distribution limit. In case of a symmetrical distribution (e.g. the normal distribution), the coefficient $\alpha_x = 0$ and only the mean μ_x and standard deviation σ_x are considered. This type of distribution is indicated in Fig. 5.5 by a full line.

The characteristic and design values of the material properties are defined as specified fractiles of the appropriate distribution. Usually the lower 5% fractile is considered for the characteristic strength and a smaller fractile probability (around 0·1%) is considered for the design value. If the normal distribution is assumed, the characteristic value x_k, defined as 5% lower fractile, is derived from the statistical parameters μ_x and σ_x as

$$x_k = \mu_x - 1{\cdot}64\sigma_x \qquad (D5.2)$$

where the coefficient $-1{\cdot}64$ corresponds to the fractile probability 5%. The statistical parameters μ_x, σ_x and the characteristic value x_k are shown in Fig. 5.5 together with the normal probability density function of the variable x (full line). The coefficient $-3{\cdot}09$ should be used when the 0·1% lower fractile (design value) is considered.

Generally, however, the probability distribution of the material property may have asymmetrical distribution with positive or negative skewness.[40,41] The dotted line in Fig. 5.5 shows the general three-parameter (one-sided) log-normal distribution having the negative coefficient of skewness $\alpha_x = -1$ and therefore the upper limit $x_0 = \mu_x + 3{\cdot}10\sigma_x$.

In the case of asymmetric distribution a fractile x_p corresponding to the probability p may be calculated from the relationship

$$x_p = \mu_x + k_{p,\alpha}\sigma_x \qquad (D5.3)$$

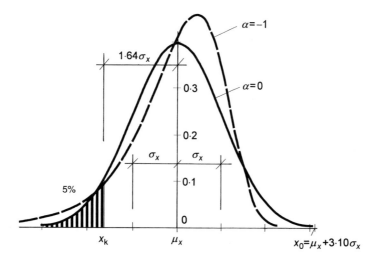

Fig. 5.5. *Characteristics of a material property*

where the coefficient $k_{p,\alpha}$ depends on the probability p and the coefficient of skewness α_x. For the three-parameter log-normal distribution, selected values of the coefficient k_α for determination of the lower 5% and 0·1% fractiles are indicated in Table 5.1.

It follows from Table 5.1 and equation (D5.3) that the lower 5% and 0·1% fractiles for the normal distribution (when $\alpha_x = 0$) may be considerably different from those corresponding to an asymmetrical log-normal distribution. When the coefficient of skewness is negative, $\alpha_x < 0$, the predicted lower fractiles for log-normal distribution are less than those obtained from the normal distribution with the same mean and standard deviation. When the coefficient of skewness is positive, $\alpha_x > 0$, the predicted lower fractiles for log-normal distribution are greater than those obtained from the normal distribution.

Therefore, when the normal distribution is used as an approximation and the correct distribution has a negative coefficient of skewness, $\alpha_x < 0$, the predicted lower fractiles will then have an unfavourable error (i.e. will be greater than the correct values). For the case when the correct distribution has a positive coefficient of skewness, $\alpha_x > 0$, the predicted lower fractiles will then have a favourable error (i.e. will be less than the correct values). In the case of the 5% lower fractile value (commonly accepted for the characteristic value) with the coefficient of skewness within the interval $< -1, 1 >$, the error created is about 6% for a coefficient of variation less than 0·2.

Considerably greater differences may occur for the 0·1% fractile value (which is approximately considered for design values), when the effect of asymmetry will be significant. For example, in the case of a negative asymmetry with $\alpha_x = -0·5$ (indicated by statistical data for strength of some grades of steel and concrete), and a coefficient of variation of 0·15 (adequate for concrete), the correct value of the 0·1% fractile corresponds to 78% of the value predicted assuming the normal distribution. When the coefficient of variation is 0·2, the correct value decreases almost to 50% of the value determined assuming the normal distribution.

However, when the material property has a distribution with a positive skewness, then the estimated lower fractile values obtained from the normal distribution may be considerably lower than the theoretically correct value corresponding to appropriate asymmetrical distribution (and therefore conservative and uneconomical). Generally, the consideration of asymmetry to determine properties is recommended (see also Appendix C to this handbook) whenever the coefficient of variation is greater than 0·1 or the coefficient of skewness is outside the interval $< -0·5, 0·5 >$. This is one of the reasons why the design value of a material property should be determined as a product of the characteristic value, which is not highly sensitive to the asymmetry, and the appropriate partial safety factor γ_m.

When the upper fractiles representing upper characteristic values are needed, equation (D5.2) may be used provided that all numerical values for the coefficient of skewness α_x and $k_{p,\alpha}$ given in Table 5.1 are taken with the opposite sign. However, in this case the experimental data should be carefully checked to avoid the possible effect of material not passing the quality test for the higher grade (see Section 5.1 and Figs. 5.1 and 5.2).

The above operational rules are applicable when the theoretical model for the probability distribution is known (e.g. from extensive experimental data or from previous experience). If, however, only limited experimental data are available, a more complicated statistical technique should be used (see Chapter 8 and Appendix C) to take account of statistical uncertainty due to limited information.

Table 5.1. The coefficient k_α for determination of the lower 5% fractile assuming three-parameter log-normal distribution

Coefficient of skewness α_x	−2·0	−1·0	−0·5	0·0	0·5	1·0	2·0
Coefficient $k_{p,\alpha}$ for $p=5\%$	−1·89	−1·85	−1·77	−1·64	−1·49	−1·34	−1·10
Coefficient $k_{p,\alpha}$ for $p=0·1\%$	−6·24	−4·70	−3·86	−3·09	−2·46	−1·99	−1·42

CHAPTER 6

Geometrical data

This chapter is concerned with geometrical data used in the design of building structures and civil engineering works. The material in this chapter is covered in *clauses 6(1)P, 6(2), 6(3) and 6(4)P in Section 6*. The appendices provide general information about the characteristics of geometrical quantities (Appendix 1) and the tolerances for overall imperfections (Appendix 2).

Clauses 6(1)P, 6(2), 6(3) and 6(4)P

6.1.   Geometrical data in design calculation

Geometrical quantities describe the shape, size and overall arrangement of structures, structural elements and cross-sections. In the design, account should be taken of the possible variation of their magnitudes, which depend on the level of workmanship in the manufacture and execution processes (setting-out, erection, etc.) on the site. In design calculations the geometrical data should be represented by the characteristic values or, in the case of imperfections, directly by their design values (*clause 6(1)P*). According to *clause 6(2)* the characteristic values usually correspond to the dimensions specified in the design, which is the nominal value (see Fig. 6.1). However, where relevant (*clause 6(3)*), values of geometrical quantities may correspond to some prescribed fractile of the statistical distribution. This value may be determined using equation (D6.1) in Appendix 1.

Clause 6(1)P,
Clause 6(2)
Clause 6(3)

According to the important principle included in *clause 6(4)P*, tolerances for connected parts shall be mutually compatible. To verify mutual compatibility of all specified tolerances, a separate analysis taking account of the imperfections and the deviations provided in Eurocodes 2 to 9 is necessary. Such an analysis is particularly important, for example, for the case when large-span precast components are used as structural elements or infill internal or external (cladding) components.[42–44] Statistical methods described in BS 5606[45] give a detailed description of methods that may be applied to verify mutual compatibility of all considered tolerances.

Clause 6(4)P

Appendix 1: Characteristics of geometrical quantities

Geometrical data are generally random variables. In comparison with actions and material properties their variability can in most cases be considered small or negligible. Such quantities can be assumed to be non-random and as specified in the design drawings (e.g. effective span, effective flange widths). However, when the deviations of certain dimensions can have a significant effect on the actions, action effects and resistance of a structure, the geometrical quantities should be considered either explicitly as random variables, or implicitly in the models for action or structural properties (e.g. unintentional

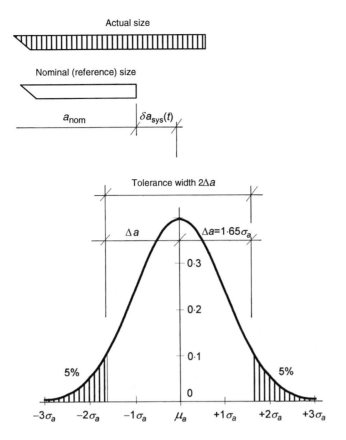

Fig. 6.1. *Characteristics of a dimension a*

eccentricities, inclinations, and curvatures affecting columns and walls). Relevant values of some geometric quantities and their deviations are usually provided in Eurocodes 2 to 9. Selected informative values for geometric quantities describing the shape, size and overall arrangement of structures are indicated in Appendix 2.

The manufacturing and the execution process (e.g. setting-out and erection) together with physical and chemical causes will generally result in deviations in the geometry of a completed structure compared with the design. Generally, two types of deviations will occur:[42,43]

(a) initial (time-independent) deviations due to loading, production, setting-out and erection
(b) time-dependent deviations due to loading and various physical or chemical causes.

The deviations due to manufacturing, setting-out and erection are also called 'induced deviations'; the time-dependent deviations due to loading and various physical and chemical causes (creep, effect of temperature and shrinkage) are called 'inherent deviations' (or deviations due to the inherent properties of structural materials).

For some building structures (particularly when large-span precast components are used), the induced and inherent deviations may be cumulative for particular components of the structure (e.g. joints and supporting lengths). The effects of the cumulative deviations in the design with regard to the reliability of the structure, including aesthetic and other functional requirements, should be taken into account.

The initial deviations of a dimension may be described by a suitable random variable and the time-dependent deviations may be described by the time-dependent systematic deviations of the dimension. To clarify these fundamental terms, Fig. 6.1 shows a probability distribution function of a structural dimension a, its nominal (reference) size a_{nom}, systematic deviation $\delta a_{sys}(t)$, limit deviation Δa and the tolerance width $2\Delta a$.

The nominal (reference) size a_{nom}, is the basic size used in design drawings and documentation, to which all deviations are related. The systematic deviation $\delta a_{sys}(t)$ is a time-dependent quantity representing the time-dependent dimensional deviations. In Fig. 6.1 the limit deviation Δa is associated with the probability 0·05, which is the probability commonly used to specify the characteristic strength. In this case the limit deviation is given as $\Delta a = 1·64\sigma_a$. In special cases, however, other probabilities may be applied and, instead of the coefficient 1·64 other values should be used. Generally, a fractile a_p of a dimension a corresponding to the probability p may be expressed as

$$a_p = a_{nom} + \delta a_{sys}(t) + k_p \sigma_a \qquad (D6.1)$$

where the coefficient k_p depends on the probability p and assumed type of distribution (see also Appendix 2 to Chapter 5, and Appendix C).

Worked example 6.1

Consider the simple assembly shown in Fig. 6.2. A prefabricated horizontal component is erected on two vertical components. The relevant dimensions describing setting-out and erection and the dimensions of the component are obvious from Fig. 6.2. For the supporting length b (which is one of the lengths b_1 and b_2 in Fig. 6.2) of the horizontal component, the nominal value $b_{nom} = 85$ mm is designed. Taking account of deviations in setting-out, manufacturing and erection it was assessed that the supporting length has approximately the normal distribution with the systematic deviation $\delta a_{sys}(t) = 0$ and the standard deviation $\sigma_b = 12$ mm. It follows from equation (D6.1) and data in Appendix C that the lower 5% fractile of the length is

$$b_{0.05} = 85 - 1·64 \times 12 = 65 \text{ mm}$$

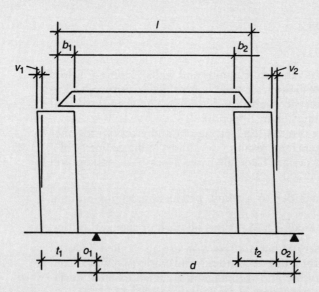

Fig. 6.2. A horizontal component erected on two vertical components

and the upper 95% fractile is

$$b_{0.95} = 85 + 1·64 \times 12 = 105 \text{ mm}$$

Therefore, with the probability 0·90 the supporting length will be within the interval 85±20 mm and the corresponding eccentricity of the loading transmitted by the component to the supporting vertical member may differ by ±10 mm from an assumed value.

The initial and time-dependent deviations may lead to a considerable variation in the shapes and sizes of structures and their parts as follows:

(*a*)   in the shape and size of cross-sections, support areas, joints, etc.
(*b*)   in the shape and size of components
(*c*)   in the overall shape and size of the structural system.

For cases (*a*) and (*b*) where the variation of structural dimensions can affect the safety, serviceability and durability of the structures, specified permitted deviations or tolerances are given in Eurocodes 2 to 9. These tolerances are denoted as normal tolerances and should be taken into account if other smaller or larger tolerances are not specified in the design. When deviating from Eurocodes 2 to 9 with regard to tolerances, the design should carefully include implications of other deviations with regard to structural reliability. For case (*c*) some informative tolerances are given in Appendix 2.

Appendix 2: Tolerances for the overall imperfections

Completed structures after erection should satisfy criteria specified in Table 6.1 and Figs 6.3–6.6.[1] Each criterion shall be considered as a separate requirement to be satisfied independently of any other tolerance criteria.

Table 6.1.   Normal tolerances after erection

Criterion	Permitted deviation
Deviation of distance between adjacent columns	± 5 mm
Inclination of column in a multi-storey building related to storey height h (see Fig. 6.3)	$0.002h$
Horizontal deviation of column location in a multi-storey building at a floor level Σh from the base, where Σh is the sum of n relevant storey heights, related to a vertical line of the intended column base location (see Fig. 6.4)	$0.0035\Sigma h/\sqrt{n}$
Inclination of a column of height h in a single-storey building other than a portal frame and not supporting a crane gantry (see Fig. 6.5)	$0.0035h$
Inclination of columns of height h in a portal frame not supporting a crane gantry (see Fig. 6.6)	Mean $0.002h$, individual $0.010h$

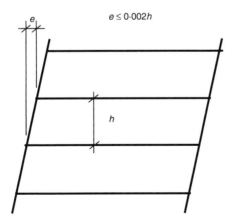

Fig. 6.3.   Inclination of a column between adjacent floor levels

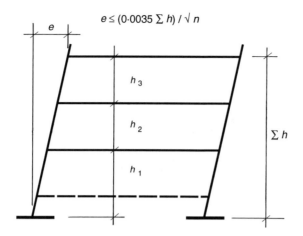

$$e \le (0{\cdot}0035 \, \Sigma \, h) / \sqrt{n}$$

Fig. 6.4. Location of a column at any floor level

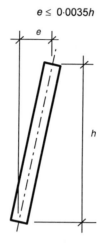

$$e \le 0{\cdot}0035h$$

Fig. 6.5. Inclination of a column in a single-storey building

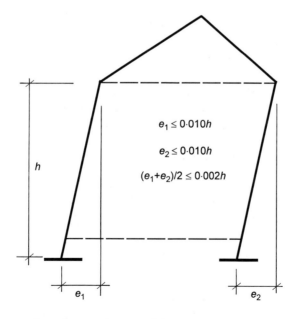

$$e_1 \le 0{\cdot}010h$$

$$e_2 \le 0{\cdot}010h$$

$$(e_1 + e_2)/2 \le 0{\cdot}002h$$

Fig. 6.6. Inclination of a column of a portal frame

CHAPTER 7

Modelling for structural analysis and resistance

This chapter is concerned with the modelling of building structures and civil engineering works for the purpose of determining action effects. The material in this chapter is covered in *Section 7*, in the following clauses:

- general   *clause 7.1*
- modelling in the case of static actions   *clause 7.2*
- modelling in the case of dynamic actions   *clause 7.3*
- modelling for fire actions   *clause 7.4*.

7.1.   General
Clause 1.3

The Principles and Application Rules presented in *Section 7* deal primarily with the modelling of structures. Action effects in cross-sections, joints or members determined through structural modelling and analysis can be used to verify the reliability of buildings and civil engineering works at their various limit states. In design calculations, the structural response to direct and indirect actions, in combination with the effects of environmental influences (e.g. interaction between fire and strength, stress and corrosion), should generally be considered.

The overall structural concept of a building or civil engineering works should be chosen with regard to the requirements of expected use, and for safety and serviceability under expected actions. It is assumed that the structural system is chosen by appropriately qualified and experienced personnel and satisfies all other requirements, e.g. concerning execution, quality control, construction materials, maintenance and use (*clause 1.3*).

Typically the structural system comprises three subsystems:[1]

(a)   the main structural system: the load-bearing elements of a building or civil engineering works and the way in which these elements function together
(b)   secondary structural elements, e.g. beams and purlins that transfer the load to the main system
(c)   other elements, e.g. cladding, roofing, partitions or façades that merely transfer loads to main and secondary elements.

While ultimate limit state failure of the main structural system can induce a global collapse with major consequences, an ultimate limit state failure of secondary or other elements would normally lead to local collapses with minor consequences. Hence, separate structural models and reliability levels may be needed for each of these subsystems or their combinations. For some structural systems, e.g. for frames with stiff infill elements such as cladding and partition walls, interaction between secondary or other elements and the main structural system should be considered.

According to *clause 7.1(1)P* the design models should involve relevant vari- *Clause 7.1(1)P* ables, and should be appropriate for predicting the structural behaviour and the limit states considered. The models should normally be based on established engineering theory and practice, verified experimentally if necessary (*clause* *Clause 7.1(2)* *7.1(2)*). Various assumptions will be necessary, such as those concerning the force–deformation or stress–strain relationships, distribution of strain in cross-sections, and adequate boundary conditions. Experimental verification may be needed, particularly for new structural systems and materials or where new theories or numerical methods are being used to analyse the structure.

Generally, any structural model should be regarded as an idealisation of the structural system. A simplified model should take account of significant factors and neglect the less important ones. The significant factors that may affect the choice of a structural model generally include the following:[1]

(*a*) geometric properties (e.g. structural configuration, spans, cross-sectional dimensions, deviations, imperfections and expected deformations)
(*b*) material properties (e.g. strength, constitutive relations, time- and stress state-dependence, plasticity, temperature- and moisture-dependence)
(*c*) actions (e.g. direct or indirect, variation in time, spatial variation, static or dynamic).

The appropriate structural model should be chosen with reference to previous experience and knowledge of structural behaviour. The sophistication of the model should take into account the intended use of results and will normally involve consideration of the appropriate limit state, the type of results and the structural response expected. Often a simple global analysis with equivalent properties can be used to identify areas that need more complex and refined modelling.

Depending on the overall structural configuration, the structure may be considered as a three-dimensional system or as a system of planar frames and/or beams. For example, a structure with no significant torsional response may perhaps be considered as consisting of a set of planar frames. Torsional response is important in structures where centres of stiffness and mass do not coincide, either by design or unintentionally (e.g. imperfections); for example, continuous flat slabs supported directly by columns and structures asymmetric in stiffness and/or mass when the three-dimensional nature of the problem needs to be considered (see Appendix 1). In many cases the form of the anticipated structural deformations due to given actions can clearly indicate a suitable simplification of the structure and therefore an appropriate structural model.[1]

When considering the stability of the structure, it is not only individual members that should be dealt with but also the whole structure: a structure properly designed with respect to individual members can still be susceptible to overall instability as a whole, e.g. torsional buckling of a uniform lattice tower crane column.

Appendix 1 demonstrates the method for the reduction process of a spatial structure.

7.2. Modelling in the case of static actions

According to *clause 7.2(1)* the modelling for static actions should normally be based on an appropriate choice of the force–deformation relationships of the members and their connections. In almost all design calculations some assumptions concerning these relationships between forces and deformations are necessary. Generally, such assumptions are dependent on the design situation, limit states and load cases being considered.

Clause 7.2(1)

The theory of plasticity which assumes the development of plastic hinges in beams, and yield lines in slabs, should be used carefully for the following reasons:

(*a*) the deformation needed to ensure plastic behaviour of the structure may not be acceptable with regard to serviceability limit states, particularly for long-span frames and continuous beams

(*b*) to avoid so-called low cyclic fatigue, these deformations should not, generally, be repeated frequently

(*c*) special attention should be paid to structures in which load carrying-capacity is limited by brittle failure or instability.

Appendix 2 provides further rules on plastic analysis.

Normally the action effects and the structural resistance are calculated separately using different design models. These models should, in principle, be mutually consistent. However, in many cases this rule may be modified in order to simplify the analysis. For example, while the analysis of a frame or continuous beam for action effects usually assumes the theory of elasticity, structural resistance of cross-sections, joints or members may be determined taking account of various non-linear and inelastic properties of the materials.

See Appendix 3 for an example of the classification of joints.

According to Eurocode 1, Part 1, the effects of displacements and deformations, i.e. second-order effects in the analysis of structures, should be considered in the context of ultimate limit state verifications (including static equilibrium) if they result in an increase of the effects of actions by more than 10% (*clause 7.2(2)*). The second-order effects may be taken into account by assuming equivalent initial imperfections and by performing a proper geometrical non-linear analysis. Alternatively, a set of additional non-linear forces, resulting from the deformations that correspond to collapse conditions, can be introduced into a first-order analysis.

Clause 7.2(2)

Generally, two types of second-order effects may be recognised:

(*a*) overall second-order effects (effects of structural sway)

(*b*) member second-order effects.

A structure may be classified as non-sway and the overall second-order effects neglected:

(*a*) if the increase of the relevant bending moments or sway shear due to second-order deformations is less than 10% of the first-order bending moment or the storey shear, respectively

(*b*) if the axial forces within the structure do not exceed 10% of the theoretical buckling load.

In the case of regular steel or concrete frames the overall analysis may be based on a first-order method and then the member analysis can take account of both overall and member second-order effects. This procedure should not be used in asymmetric or unusual cases.

Design models used to verify serviceability limit states and fatigue are usually based on the linear-elastic behaviour of the structural material (*clause 7.2(3)*), while those used to verify the ultimate limit states often take account of non-linear properties and post-critical behaviour of structures. However, in some cases of non-linear or time-dependent structural properties, for example when plastics or timber are used as structural materials, linear elastic theory may be insufficient to verify the serviceability limit states.

Clause 7.2(3)

When phenomena such as differential settlement and whole structure stability may affect the structural design under static conditions, the modelling of a structure should also include the consideration of the foundation, and this includes the soil. Soil–structure interaction can be addressed through either a simultaneous analysis of the soil structure system or a separate analysis of each system.

Appendix 4 provides further rules and practical examples for considering soil–structure interaction.

Soil–structure interaction might also have a dynamic nature (see Section 7.3).

7.3.   Modelling in the case of dynamic actions

For the limit state verification, most of the actions as specified in Eurocode 1, Part 1 are transformed into equivalent static actions. These equivalent forces are defined such that their effects are the same as, or similar to, the effects of the actual dynamic actions. When dynamic actions are simulated by quasi-static actions, the dynamic parts are considered either by including them in the static values or by applying equivalent dynamic amplification factors to the static actions (*clause 7.3(1)*). For some equivalent dynamic amplification factors, the natural frequencies have to be determined.

Clause 7.3(1)

The models for dynamic analyses of structures, to be used in the determination of the action effects or equivalent static actions, should be established such that all relevant structural elements and their mechanical properties including mass and damping ratios are realistically considered. For most cases, if the dynamic actions are caused by the motion of masses that are themselves being supported by the structure (e.g. people, machinery, vehicles), these masses should be considered in the analysis as they may affect the dynamic properties of lightweight systems.

In the case of dynamic effects caused by significant soil–structure interaction (e.g. the transmission of traffic vibrations to buildings), the contribution of the soil may be modelled, for example, by appropriate equivalent springs and dampers. The soil may also be approximated with a discrete model.

In some cases (e.g. for cross-wind vibration or seismic actions) the actions are significantly controlled by the structural response and can therefore be defined only when the dynamic behaviour of the structure has been estimated by using a modal analysis. A linear material and linear geometric behaviour may usually be accepted (*clause 7.3(2)*) for such a modal analysis. Non-linear material properties can be approximated using iterative methods together with the secant stiffness as related to the response level. For structures where only the fundamental mode is relevant, an explicit modal analysis may be replaced by an analysis with equivalent static actions, depending on mode shape, natural frequency and damping. Structural response to dynamic actions may be determined in terms of time histories or in the frequency domain (*clause 7.3(3)*).

Clause 7.3(2)

Clause 7.3(3)

Particular attention should be given to dynamic actions which may cause vibration that may infringe serviceability limit states. Guidance for assessing these limit states is given in *Annex C*.

7.4. Modelling for fire actions

Clause 7.4 provides only basic Principles and Application Rules for modelling of thermal actions on structures exposed to fire. A more detailed description of thermal actions due to fire and modelling for the structural design of buildings and civil engineering works is provided in Part 2.2 of Eurocode 1 and relevant parts of Eurocodes 2–6 and 9. As stated in *clause 2.3(2)P*, the thermal actions on structures from fire are classified as accidental actions and considered in the accidental design situation (see also *Section 9*, Eurocode 1, Part 2.2 and Chapter 9 of this handbook).

Clause 7.4

Clause 2.3(2)P

According to *clause 7.4(1)P*, appropriate models for the fire situation, which shall be used to perform the structural analysis for fire design, should involve three fundamental aspects characterising the appropriate accidental situation:

Clause 7.4(1)P

(*a*) thermal actions
(*b*) mechanical actions
(*c*) structural behaviour at elevated temperatures.

Thermal and mechanical actions to be considered in the fire situation are specified in Eurocode 1, Part 2.2, where two procedures are distinguished (*clause 7.4(2)*):

Clause 7.4(2)

(*a*) nominal temperature–time curves
(*b*) parametric temperature–time curves.

In most cases thermal actions on structures exposed to fire are given in terms of nominal temperature–time curves. These curves are applied for the specified period for which the structure is designed, using prescriptive rules or calculation models. Appropriate data for these actions are given in the main text of Eurocode 1, Part 2.2.

Parametric temperature–time curves are calculated on the basis of physical parameters for which structures are designed using calculation models. Some data and models for physically based thermal actions are given in the informative annexes of Eurocode 1, Part 2.2.

If they are likely to act in the fire situation, direct and indirect actions should be considered as for normal temperature design. *Annex F* of Eurocode 1, Part 2.2 provides guidance regarding the simultaneity of actions and combination rules applicable to structures exposed to fire. This guidance concerns permanent actions, variable actions and additional actions due to the collapse of structural elements and heavy machinery. For the special cases, where indirect actions need not be considered, simplified rules for combining of actions are also indicated in Eurocode 1, Part 2.2.

Thermal and structural models for various construction materials, given in Eurocodes 2–6 and 9, should be used to analyse the structural behaviour at elevated temperatures. *Clause 7.4(3)* gives some simplified rules and assumptions concerning the thermal models and structural models in fire exposure, which may be used when relevant to the specific material and the method of assessment. Further, the same clause allows consideration of the behaviour of materials or sections at elevated temperatures as linear-elastic, rigid-plastic or non-linear.

Clause 7.4(3)

At present a number of experimental investigations are being carried out to support more advanced thermal and structural models for fire actions. Tabulated data given in Eurocodes 2–6 and 9 have mainly been obtained from test results or numerical simulations based only on the action as described by the standard fire exposure (*clause 7.4(4)*).

Clause 7.4(4)

Appendix 1: Modelling of three-dimensional framed structures for static or quasi-static analysis

Modelling of a three-dimensional structure can generally be carried out in three steps as follows.

(a)   At first, the three-dimensional structure (Fig 7.1) can in general be considered as an assembly of members.[1] These constituent parts can be plane sub-assemblies such as floor structures, individual frames consisting of linear ties, struts, beams and columns, and even two- or three-dimensional elements such as slabs, walls or box-type core elements. The idealisation of the structure should also include proper support (boundary) conditions. The modelling should be such that the effects of actions on the actual structure is properly assessed. In particular, stability aspects, performance of joints, and foundation conditions should be adequately considered.

(b)   As the second step, the individual plane frames may, in general, be analysed only against in-plane actions, using in-plane resistances. That is, out-of-plane effects of such actions and imperfections may be neglected in this analysis.

(c)   In the third step, member verifications can be carried out by isolating them from the frame, and applying all internal and external forces, restraints and out-of-plane imperfections.

When the horizontal loads are distributed to individual frames through stiff horizontal diaphragms (e.g. floor slabs, horizontal bracings), this should be done in relation to the centre of stiffness of the frames and according to their respective stiffnesses (Fig. 7.2). In doing this, the torsional effect of any eccentricity between the centre of stiffness and the

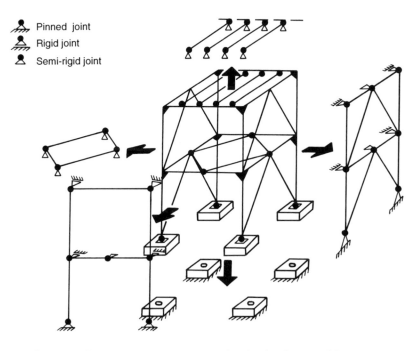

Fig 7.1.   Reduction of a spatial structure to individual sub-assemblies

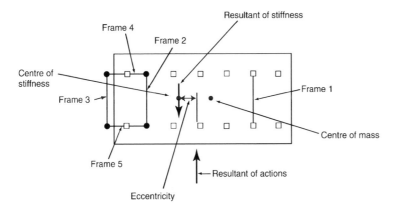

Fig. 7.2. Distribution of horizontal actions in the presence of stiff horizontal diaphragms

resultant of the actions should also be considered. In finding the centre of stiffness, frames with low in-plane stiffness may not be considered.

Appendix 2: Special rules for plastic analysis

For the first step in plastic analysis, a plastic hinge mechanism is assumed to develop at the ultimate limit state of the structure. The moment distribution is then determined on the basis of the plastic strength (e.g. ultimate plastic bending moments) of members and joints assuming sufficient rotation capacity of the hinges (see Figs 7.3 and 7.4).

Sufficient rotational capacity can be assured by verifying that the rotations in the plastic hinges φ_{Ed} at the ultimate limit state fulfil

$$\varphi_{Ed} < \varphi_{Rd}/\gamma_\varphi \qquad (D7.1)$$

where φ_{Rd} is rotational capacity and γ_φ is the safety factor.

The distribution of the horizontal loads to frames may be performed according to the plastic resistance of the frames instead of the stiffness of the frames. Such a distribution is in accordance with the assumption for compatible deflections of the frames.

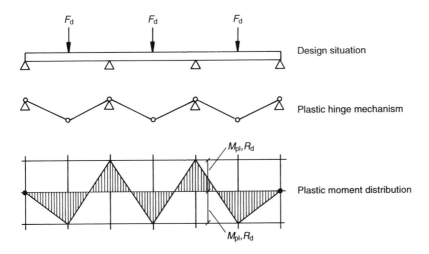

Fig. 7.3. Formation of a plastic hinge mechanism

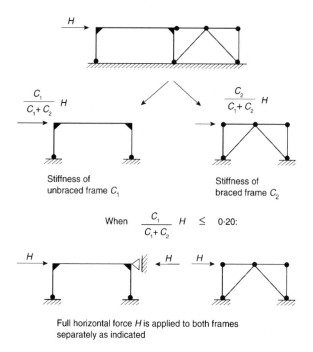

Fig. 7.4. Distribution of horizontal forces to frames with different stiffness

Appendix 3: Classification of joints

In the analysis of frames, it may be necessary to classify joints according to their rigidities. Although the following example provides such a classification for structural steel joints, the same principles can be used with respect to joints in reinforced concrete structures.

Figure 7.5 provides a classification of steel joints based on their rigidities. In Fig. 7.5, S is the idealised joint stiffness, ζ is the joint rotation, ζ_{Rd} is the design rotation, M is the applied moment, and M_{Rd} is the design resistance (moment).

A rigid connection can be defined as one whose deformation does not affect the moment distribution by more than 5% of that under the 'rigid' assumption. A hinged connection can be defined as one with negligible stiffness, and where rotations will be either within the elastic limits or in the plastic range, without giving rise to significant restraints. In the case of semi-rigid connections, which may be modelled by representative springs, the performance is between the above two extremes. (Further details on these aspects are available in Eurocodes 2 to 4.)

Appendix 4: Soil–structure interaction

The following terminology is assumed for this appendix (see Fig. 7.6):

- *soil* describes the supporting system of the ground, responding to pressure from the foundation members
- *foundation* includes soil plus foundation members
- *structure* includes the foundation members plus whatever they support (e.g. the building)
- *superstructure* denotes the building frames etc.

The following two examples are about the consideration of soil–structure interaction under static conditions in buildings with spread foundations.

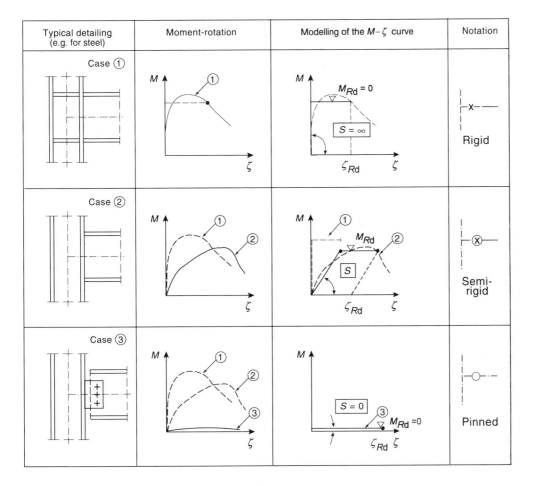

Typical detailing (e.g. for steel)	Moment-rotation	Modelling of the $M-\zeta$ curve	Notation
Case ①		$M_{Rd} = 0$ $S = \infty$	Rigid
Case ②	① ②	① M_{Rd} ② S	Semi-rigid
Case ③	① ② ③	$S = 0$ ③ $M_{Rd} = 0$	Pinned

Fig. 7.5. *Classification of structural steel joints*
In rows 2 and 3 of column 2, the M–ζ curves for preceding cases are shown as dashed lines. In column 3, the idealised curves are shown as solid lines and the experimental curves are shown as dashed lines

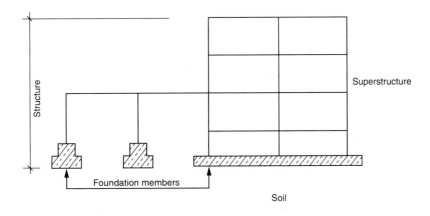

Fig. 7.6. *Terminology*

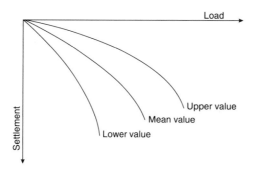

Fig. 7.7. Load–settlement behaviour of soils under different properties

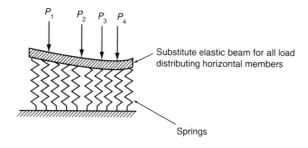

Fig. 7.8. Example of soil modelling

Simultaneous analysis

The stiffness and load-bearing capacity of the soil may be derived from the possible load–settlement behaviour, an example of which is shown in Fig. 7.7. These properties can be used in the analysis of the soil–structure system, for example, as shown in Fig. 7.8. In such a model, the soil may be modelled either as a set of (coupled) springs or as an elasto-plastic continuum.

The analysis, while linear for serviceability limit state verifications, may be non-linear for ultimate limit state verifications. This non-linearity is with respect to both the structure and the soil. In non-linear analysis, under small load increments, the development of forces and moments should be carefully observed in relation to global stability of the building and bearing capacities of the superstructure, the foundation members and the soil.

If the analysis is based on the mean properties X_m, the design value P_d can be obtained as[46]

$$P_d = P_{ult} \frac{1}{\gamma_M} \frac{X_k}{X_m} \tag{D7.2}$$

where P_{ult} is the resistance under the mean value, and X_k and γ_M are the characteristic (lower) value of the soil property and the partial safety factor respectively.

Separate analysis of systems

In this case, the soil is first considered not to be deformable (Fig. 7.9). The internal forces at the soil–structure interface are calculated for the characteristic loads (rare combination). Then, the soil deformations due to these forces are calculated and, in turn, used as characteristic values for imposed settlements acting on the structure (Fig. 7.10).

If the design of the foundation structure and the structural frame are sensitive to differential settlements, then this process should be carried out iteratively by repeating the above two steps. For an elastic analysis, the iterations may be stopped when the error is less than 5% of the total action effects in the frame.

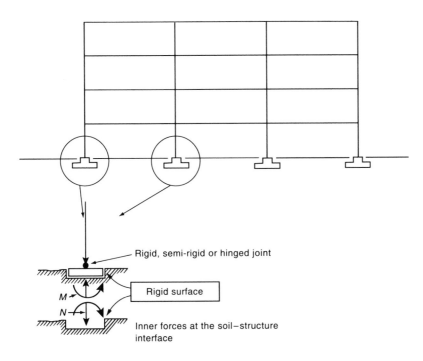

Fig. 7.9.   *First step assumption for soil behaviour*

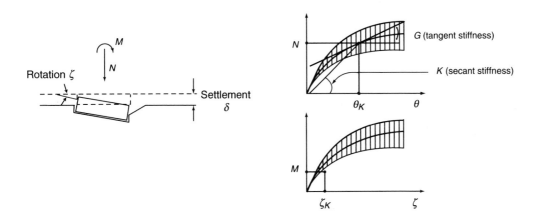

Fig. 7.10.   *Application of foundation settlements as imposed frame deformations*

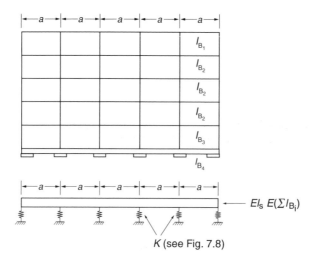

Fig. 7.11.   Substitute beam model of foundation system for soil–structure interaction

If the frame is regular, and if a beam on elastic springs (see Fig. 7.11) can be considered as an appropriate model for the foundation analysis, then the 5% error criterion can be expressed as

$$EI_S \leq 0 \cdot 002KL^3 \qquad\qquad (D7.3)$$

as the condition for which no further step, *i.e.* no consideration of soil–structure inter-action, is necessary. K is the stiffness of the linear-elastic (soil) spring. If a non-linear spring is assumed, K is the secant stiffness that will need to be approximated. EI_S is the bending stiffness of the substitute beam (see Fig. 7.11). The above condition has been derived by considering the change of forces due to the settlement of a single support while other supports remain rigid. No beam rotation is considered to occur at the supports, and the beam spans adjoining the particular support are considered equal to *a*.

CHAPTER 8

Design assisted by testing

This chapter is concerned with the design assisted by testing of building and civil engineering structures. The material in this chapter is covered in *Section 8*, in the following clauses:

- general   *clause 8.1*
- types of tests   *clause 8.2*
- derivation of design values   *clause 8.3*.

In addition to the main text of *Section 8*, this chapter covers *Annex D*, which gives guidance for the planning and evaluation of tests to be carried out in connection with structural design, where the number of tests is sufficient for a meaningful statistical interpretation of their results. Basic statistical techniques for estimating fractiles are briefly described in references 1 and 47 and in Appendix C. Some of the procedures described in this chapter may also be useful for the assessment of existing structures.

8.1. General

Design assisted by testing is a procedure using physical testing (e.g. models, prototypes, in situ) for establishing design values. Such procedures can be used in particular for cases where the calculation rules or material properties given in Eurocodes 1 to 9 are considered insufficient, or where a more economical design may result (*clause 8.1(1)P*).

Clause 8.1(1)P

As stated in *Annex D*, testing might, for example, be carried out under the following circumstances:

(a) if material properties or load parameters are not adequately known, e.g. in cases that cannot be treated by using the guidance given in Eurocodes 1 to 9 because of the lack of adequate theoretical models or data

(b) when adequate calculation models are not available or the design procedures seem to lead to uneconomical results

(c) if a large number of similar components will be used and a significant reduction in cost can be expected

(d) if the structural behaviour is of special concern, e.g. when deformation under service conditions may be critical

(e) to define the control checks that are needed to give reassurance about the proper behaviour of the structure.

An essential requirement for using design assisted by testing is given in *clause 8.1(2)P*. Tests should be set up and evaluated in such a way that the structure has

Clause 8.1(2)P

the same level of reliability, with respect to all possible limit states and design situations, as would be achieved by design using Eurocodes 1 to 9. Hence, all uncertainties, such as those due to the conversion of experimental results and those arising from the statistical uncertainty associated with design assisted by testing (e.g. the application of experimental results, and any statistical uncertainties due to the sample size) should be taken into account. Furthermore, the conditions during testing should so far as possible be representative (*clause 8.1(3)*) of those that can be expected to arise in practice.

Clause 8.1(3)

Clause 8.1(4) is not clear as it stands. All design methods involve simplifications and this may need to be allowed for when analysing test results. For example the interpretation of results obtained from tests on reinforced concrete beams might have to allow for the tensile strength of the concrete which is normally ignored in design guidance (e.g. in Eurocode 2).

Clause 8.1(4)

8.2. Types of tests

Prior to the execution of any tests, a test plan should be agreed with all interested parties including the testing organisation. This plan should state the objectives of the tests and all specifications necessary for the selection or production of the test specimens, the execution of the tests and, if appropriate, the methods of analysis. With regard to the objectives of the tests, several types of test are clearly defined in *clause 8.2(1)*. Some types of test are carried out before the design in order to obtain design values (test types (*a*), (*b*) and (*c*) in *clause 8.2(1)*). Where test results will not be available at the time of design (e.g. tests (*d*), (*e*) and (*f*) from *clause 8.2(1)*), the intention behind *clause 8.2(2)* is far from clear. What it seems to be trying to say is that if control tests are going to be carried out some time in the future, then the design procedure should meanwhile continue on the assumption that those tests will be passed. For example, if 40 MPa concrete is specified in the design, then the design process should continue on the assumption that concrete meeting this specification will actually be provided and confirmed by cube or cylinder testing.

Clause 8.2(1)

Clause 8.2(2)

8.3. Derivation of design values

The derivation of the design values (e.g. for a material property, a model parameter or a resistance) from results of tests should normally be carried out by appropriate statistical techniques, taking account of the sample size. According to *clause 8.3(1)P* the design value may be derived in either of the following ways:

Clause 8.3(1)P

(*a*) by assessing a characteristic value, which may have to be modified by conversion factors, and by applying partial factors
(*b*) by determining the design value directly from test results allowing for conversion aspects.

According to method (*a*), a suitable statistical technique is first used to assess a characteristic value; this is then divided by a partial factor and possibly multiplied by conversion factors to derive the design value. According to method (*b*), the design value is assessed directly from tests, allowing implicitly or explicitly for conversion aspects.

Both methods (*a*) and (*b*) are based on the application of statistical techniques to assess a certain fractile from a limited number of test results (see Appendix C). For method (*a*), usually the 5% fractile is assumed for the characteristic value, while for method (b) much lower probabilities, about 0.1% will be appropriate. Normally method (b) would involve many more tests and so will be much more expensive than method (a).

Any assessment of probabilities involves basically two types of uncertainty, which are to a certain extent related:

(i) statistical uncertainty due to the limited sample size
(ii) uncertainty due to vague prior information about the nature of statistical distribution.

These uncertainties will give rise to a considerable error which will usually be much greater for fractiles corresponding to a low probability (0.1%) than for the 5% fractiles. This is the principal reason why, in general, method (a) will usually be used (*clause 8.3(2)*). Direct assessment of the design value, method (b), should be used only where a sufficiently large number of test results are available together with substantial information supporting the appropriate statistical model.

Clause 8.3(2)

When method (a) is used, consideration should always be given to both types of uncertainty, together with conversion factors (*clause 8.3(2)*) resulting from influences not sufficiently covered by the tests, such as:

Clause 8.3(2)

(a) size effects (often, tests are performed using smaller specimens than actual elements)
(b) time effects (normally, tests are performed only under short-term loading)
(c) boundary conditions (e.g. free, rigid, semi-rigid, effect of joints and remaining parts)
(d) temperature and humidity conditions (e.g. effects on material strength).

In addition, with regard to workmanship when the conditions during testing differ significantly from those expected for the intended structure in its actual environment this should be allowed for.

Clause 8.3(3) gives some guidance for using method (b). This guidance, in the opinion of the authors, may be insufficient for practical applications, and the user is advised to obtain specialist advice. Some guidance concerning the required level of reliability is given in *Annex A*. Information about statistical uncertainties and appropriate statistical techniques is given in Appendix C.

Clause 8.3(3)

Further information about strength of specific materials or resistance of structural components may be found in Eurocodes 2 to 9 (*clause 8.3(4)*). Additional information concerning statistical techniques is available in Appendix C and *Annex D*.

Clause 8.3(4)

CHAPTER 9

Verification by the partial factor method

This chapter is concerned with the verification of building structures and civil engineering works using the partial factor method. The material in this chapter is covered in *Section 9*, in the following clauses:

- general   *clause 9.1*
- limitations and simplifications   *clause 9.2*
- design values   *clause 9.3*
- ultimate limit states   *clause 9.4*
- serviceability limit states   *clause 9.5*.

9.1. General

In the Eurocodes, the reliability of buildings and civil engineering works is based on the concept of limit states and their verification by the partial factor method. In this method a structure is considered to be reliable if, for all the relevant design situations, the limit states are not exceeded when the design values of the basic variables (the actions, material properties and geometrical data) are used in the design models (*clause 9.1(1)P*).

Clause 9.1(1)P

For the verification of structural reliability the following essential elements are always involved:

(*A*) the various design models as follows:

  (*a*) for verification of the ultimate limit states
    (i) the consideration of overall stability, where it should be verified that the design effects of destabilising actions are less than the design effects of stabilising actions
    (ii) the consideration of rupture or excessive deformation of a section, member or connection, where it should be verified that the design effects expressed in terms of internal forces or moments are less than their resistance values
    (iii) the consideration of overall structural behaviour, where it should be verified that the structure is not transformed into a mechanism
  (*b*) for verification of the serviceability limit states, where it should be verified that the design effects of actions do not exceed serviceability constraints

(*B*) the design values of the appropriate basic variables (the actions, the material properties and geometrical data) which are derived from their characteristic values, and a set of partial factors, and combination and other coefficients. In some cases, e.g. when the design procedure is performed on the basis of tests, the design value may be determined directly from available statistical data (more details are available in Chapter 8, *Section 8* and *Annex D*.

Clause 9.1(2)P states that verification shall in particular be carried out at the ultimate limit states (where the effects of design actions do not exceed the design resistance of a structure) and the serviceability limit states (where the effects of design actions do not exceed the performance criteria). In addition, other criteria and verifications may also need to be considered (e.g. verification for fatigue, consideration of the deterioration of the structural materials to be used, detailing according to Eurocodes 2 to 9, and good practice). The limit states should be verified for all decisive (critical) design situations and load cases. As stated in *clause 2.3(1)P* (see also Chapter 2), 'the selected design situations shall be sufficiently severe so as to encompass all conditions which can reasonably be foreseen to occur during the execution and use of the structure'.

Clause 9.1(2)P

Clause 2.3(1)P

Eurocode 1, Part 1 requires the consideration of selected design situations and identification of critical load cases. For each of the critical load cases, the design values of the effect of actions in combination (*clause 9.1(3)*) should be determined. A load case identifies compatible load arrangements, sets of deformations and imperfections which should be considered simultaneously for a particular verification (*clause 9.1(4)*). Actions that cannot occur simultaneously for physical reasons should not be considered together in a combination (*clause 9.1(5)*). However, to simplify design calculations some incompatible actions are sometimes considered together even though they cannot, for physical reasons, occur simultaneously, e.g. concentrated and uniformly distributed traffic actions as defined in Eurocode 1, Part 3.

Clause 9.1(3)

Clause 9.1(4)
Clause 9.1(5)

General rules for different load arrangements, identifying the position, magnitude and direction of a free action, are given in Eurocode 1, Parts 2, 3 and 4 (*clause 9.1(6)*). Possible deviations from assumed directions or positions of free actions should be considered in design calculations (*clause 9.1(7)*). The design values of the relevant basic variables (the actions, the material properties and geometrical data) are generally dependent on the limit state and load case considered, as described later in this chapter. Consequently, several design values may be needed for some basic variables in reliability verifications of a structure (*clause 9.1(8)*).

Clause 9.1(6)
Clause 9.1(7)

Clause 9.1(8)

Considering *Table 9.2*, for verification at the ultimate limit states different design values for actions are used:

(*a*) for the consideration of static equilibrium (case A)
(*b*) for the consideration of failure of structural elements (case B)
(*c*) for the consideration of failure in the ground (case C).

In addition, different design values are needed for verification of the serviceability limit states.

9.2. Limitations and simplifications

The Principles and Application Rules in Eurocode 1, Part 1 are limited to ultimate limit states and serviceability limit states for the structures subjected to static loading, and quasi-static loading where the dynamic effects are assessed using equivalent static loads and dynamic amplification factors, e.g. in the case of wind load. Eurocode 1, Part 1 does not completely cover non-linear and dynamic analysis and fatigue (*clause 9.2(1)*), but modifications for non-linear analysis and fatigue are provided in other parts of Eurocode 1 and in Eurocodes 2 to 9.

Clause 9.2(1)

Simplified verification based on the limit states concept in connection with the partial factor method is provided in Eurocode 1, Part 1 (see *clauses 9.4.5* and *9.5.5* and the Sections in this chapter on 'Simplified verification for building structures'), and may be used

Clauses 9.4.5 and 9.5.5

(*a*) only for limit states and load combinations that are known from previous experience to be potentially critical

(*b*) by specifying particular detailing rules and other provisions with regard to the production and execution processes to meet all the safety and serviceability requirements (*clause 9.2(2)*).

Clause 9.2(2)

9.3. Design values

In the structural Eurocodes the appropriate basic variables (the actions, the material properties and the geometrical data) are derived from their characteristic or representative values, and the design values are determined using these values together with a set of partial factors, and combination and other coefficients. The statistical techniques to assess the design values directly using available data, provided in Appendix C, *Section 8* and *Annex D*, should be limited to cases where sufficient experimental evidence and reliable previous experience are available. For determining the design values from the characteristic values, Eurocode 1, Part 1 specifies Principles and Application Rules separately for actions, action effects, materials properties, geometric data and resistances.

Design values of actions

In accordance with *clause 9.3.1(1)P*, the design value F_d of action F is generally determined from its representative value F_{rep} as

Clause 9.3.1(1)P

$$F_d = \gamma_F F_{rep} \qquad (9.1)$$

where γ_F is the partial factor and F_{rep} is the representative value of the action (see Section 4.3). The partial factor for the action γ_F should take account of the possibilities of unfavourable deviations of the actions, inaccuracies of the action models and uncertainties in the assessment of actions. Thus, the factor for the action γ_F is in most cases dependent on various deviations and uncertainties which will have a different statistical distribution or may be very difficult to describe by a statistical model at all (e.g. uncertainties in the assessment of action for which a statistical model is not available). Consequently the partial factors recommended in the present generation of structural Eurocodes are mainly derived from previous experience, and not solely using theories of reliability.

F_{rep} is the representative value of an action F where for permanent actions

$$F_{rep} = G_k \qquad (D9.1)$$

and for variable actions

$$F_{rep} = \psi_i G_k \qquad (D9.2)$$

where for the dominant variable actions in the verification expressions provided in Eurocode 1, Part 1 and this chapter $\psi_i = 1$, and for non-dominant actions $\psi_i = \psi_0$, ψ_1 and ψ_2 (see also Chapter 4).

Thus, depending on the type of verification and combination procedures, design values for particular actions are expressed as follows in Eurocode 1, Part 1 (*clause 9.3.1(2)*)

Clause 9.3.1(2)

$$G_d = \gamma_G G_k \text{ or } G_k$$
$$Q_d = \gamma_Q Q_k, \gamma_Q \psi_0 Q_k, \psi_1 Q_k, \psi_2 Q_k \text{ or } Q_k$$
$$A_d = \gamma_A A_k \text{ or } A_d \qquad (9.2)$$
$$P_d = \gamma_P P_k \text{ or } P_k$$
$$A_{Ed} = A_{Ed}$$

These equations represent special forms of general expressions (D9.1) and (D9.2). The alternatives indicated in the first four equations indicate possible forms of expressions (D9.1) and (D9.2) used in the verification of different limit states and load combinations. For example, the design value of a permanent action for the verification of the ultimate limit states is given as $\gamma_G G_k$, and for verification of the serviceability limit states as G_k. Moreover, as indicated in *clause 9.1.3(3)P*, where distinction has to be made between favourable and unfavourable effects of permanent actions, two different partial factors shall be used. Further, for seismic actions the design value may depend on the structural behaviour characteristics (*clause 9.1.3(4)*, see also Eurocode 8).

Clause 9.1.3(3)P

Clause 9.1.3(4)

Design values of the effects of actions

The effects of actions E are responses (for example internal forces and moments, stresses, strains and displacements) of the structure to the actions imposed on the structure. The effects of actions E are dependent on the actions F, the geometrical properties a and material properties X and may be expressed mathematically as in the functional expression

$$E = \text{function}(F_1, F_2, \ldots a_1, a_2, \ldots X_1, X_2, \ldots) \qquad (D9.3)$$

A simple operational rule to determine the design value R_d is provided in *clause 9.3.2(1)P*. For a specific load case the design value E_d of the effect of actions E is determined from the design values of the actions F, geometrical data a and material properties X when relevant as follows

Clause 9.3.2(1)P

$$E_d = E(F_{d1}, F_{d2}, \ldots a_{d1}, a_{d2}, \ldots X_{d1}, X_{d2}, \ldots) \qquad (9.3)$$

where 'function' in (D9.3) is represented by E and the design values $F_{d1}, \ldots$, $a_{d1}, \ldots$ and $X_{d1}, \ldots$ of basic variables are chosen according to *clauses 9.3.1, 9.3.3 and 9.3.4* respectively.

Clauses 9.3.1, 9.3.3 and 9.3.4

Equation (9.3) represents a simple operational rule which does not directly imply any probabilistic property of E_d and should be considered as a useful simplification only. In addition there may be some uncertainties involved in the model (D9.3) itself. For example, in some more complicated cases, in particular for non-linear analysis or for modelling of structures with unusual geometry (e.g. complicated shell roofs), the effect of the uncertainties in the models used in the calculations should be considered explicitly (*clause 9.3.2(2)*). This leads to the application of a coefficient of model uncertainty γ_{Sd}, applied either to the actions or to the action effects, whichever is the more conservative. The factor γ_{Sd} may refer to uncertainties in the action model and/or the action effect model.

Clause 9.3.2(2)

Specific problems, using the partial factor method, may arise in the case of non-linear analysis, i.e. when the action effect is not proportional to the action, the partial safety factors should be applied with caution and a general reliability analysis as indicated in *Annex A* should be considered. *Clause 9.3.2(3)*, however, provides some approximate rules for non-linear analysis which may be considered if there is a single predominant action. For this case the partial factor is applied either to the representative value of the action or to the action effect corresponding to the representative value of the action, whichever is more conservative.

Clause 9.3.2(3)

Design values of material properties

The design value X_d of a material or a product property X is generally determined directly from the characteristic value X_k using the partial factor for the material or product property γ_M and, if relevant, the conversion factor η in accordance with the relationship

$$X_d = \eta X_k / \gamma_M \quad \text{or} \quad X_d = X_k / \gamma_M \qquad (9.4)$$

The conversion factor η takes account of the effect of the duration of the load, volume and scale effects, effects of moisture and temperature, etc. In some cases the conversion is implicitly considered when determining the characteristic value itself, as indicated by the definition of η, or by γ_M.

The partial factor γ_M, which is given in Eurocodes 2 to 5, should cover unfavourable deviations from the characteristic values, inaccuracies in the conversion factors, and uncertainties in the geometric properties and the resistance model.

Design values of geometrical data

Variability of geometrical quantities is usually less significant than variability of actions and material properties, and in many cases is negligible (see Chapter 6). Consequently, design values of geometrical data are generally represented by the nominal values (*clause 9.3.4(1)P*)

Clause 9.3.4(1)P

$$a_d = a_{nom} \qquad (9.5)$$

The nominal (reference) size a_{nom} is the basic size used in design drawings and documentation, to which all deviations of the geometric quantities are related (for more details see Chapter 6). Where necessary, Eurocodes 2 to 9 may give further specifications.

In some cases when deviations in the geometrical data have a significant effect on the reliability of a structure (e.g. slender columns or thin silo walls), the geometrical design values are defined by (*clause 9.3.4(2)P*)

Clause 9.3.4(2)P

$$a_d = a_{nom} + \Delta a \qquad (9.6)$$

where Δa takes account of the possibility of unfavourable deviations from the characteristic values (for more details for determining Δa see Chapter 6). The characteristic value of a geometric quantity is usually equal to the nominal value (see *clause 6(2)*). However, Δa is introduced only where the influence of deviations is critical, e.g. imperfections in buckling analysis. Values of Δa are given in Eurocodes 2 to 9.

Clause 6(2)

Design resistance

The model for the structural resistance may be represented similarly as for action effect. The resistance R is generally dependent on the geometrical data a and on the material properties X and may be expressed mathematically as the functional relationship

$$R = \text{function}(a_1, a_2, \ldots X_1, X_2, \ldots) \qquad \text{(D9.4)}$$

A simple operational rule to determine the design value R_d is provided in *clause 9.3.5(1)P*. For a specific load case the design value R_d of the resistance R is determined from the design values of the geometrical data a and material properties X from the expression

Clause 9.3.5(1)P

$$R_d = R(a_{d1}, a_{d2}, \ldots X_{d1}, X_{d2}, \ldots) \qquad \text{(9.7)}$$

where 'function' in (D9.4) is represented by R, and $a_{d1}, \ldots$ is defined in *clause 9.3.4* and $X_{d1}, \ldots$ in *clause 9.3.3*.

Clause 9.3.4
Clause 9.3.3

Operational verification formulae introduced in Eurocodes 2 to 6 based on the principle of expression (9.7) differ and may have one of the following forms (see *clause 9.3.5(2)*)

Clause 9.3.5(2)

$$R_d = R(X_k/\gamma_M, a_{nom}) \qquad \text{(9.7a)}$$

for Eurocodes 2 and 5

$$R_d = R(X_k, a_{nom})/\gamma_R \qquad \text{(9.7b)}$$

for Eurocode 3

$$R_d = R(X_k/\gamma_m, a_{nom})/\gamma_{rd} \qquad \text{(9.7c)}$$

for Eurocode 4, where γ_M is a partial factor for a material property, also accounting for model uncertainties and dimensional variation, γ_R is a partial factor for the resistance, including uncertainties in material properties, model uncertainties and dimensional variation, γ_m is a material factor, and γ_{rd} is a factor covering uncertainties in the resistance model and in the geometrical properties.

Both partial factors γ_M and γ_R account for uncertainties of a material property (or resistance), for model uncertainties and dimensional variation.

The difference between equation (9.7a) for Eurocode 2 and equation (9.7b) for Eurocode 3 can be explained by considering bending resistance, where for a concrete beam two material factors (called γ_M in Eurocode 2) for concrete and reinforcement need to be applied, while for a steel beam one (called γ_R in Eurocode 3) needs to be applied. The last expression (9.7c) for composite structures (Eurocode 4) takes account of previous equation (9.7a) for Eurocode 2, equation (9.7b) for Eurocode 3, and the relationship $\gamma_M = \gamma_m\gamma_{rd}$.

It is shown in *Annex A* that these factors may be expressed as

$$\gamma_M = \gamma_m\gamma_{Rd}/(1 + \Delta a/a_{nom}) \qquad \text{(A.15)}$$

for Eurocodes 2 and 5

$$\gamma_R = \gamma_m\gamma_{Rd}/(1 + \Delta a/a_{nom}) \qquad \text{(A.16)}$$

for Eurocode 3

$$\gamma_{rd} = \gamma_{Rd}/(1 + \Delta a/a_{nom}) \qquad \text{(A.17)}$$

for Eurocode 4. It follows from these equations that

$$\gamma_M = \gamma_R = \gamma_m\gamma_{rd} \qquad \text{(D9.5)}$$

which may also help to clarify differences between equations (9.7a), (9.7b) and (9.7c) (see also the Appendix to this chapter). For non-linear models, or in the case of multi-variable load or resistance models, commonly encountered in Eurocodes, these relationships become more complicated.

The design resistance may also be obtained directly from the characteristic value of a product resistance, without explicit determination of design values for individual basic variables (*clause 9.3.5(3)*), from

Clause 9.3.5(3)

$$R_d = R_k / \gamma_R \qquad (9.7d)$$

This is applicable for steel members, piles, etc., and is often used in connection with design by testing (see also Chapter 8, and *Section 8* and *Annex D*).

9.4. Ultimate limit states

In Eurocode 1, Part 1 three basic cases for ultimate limit state verifications need to be considered separately as relevant. These three cases are:

(*a*) loss of equilibrium where strength of material and/or ground is insignificant (case A)

(*b*) failure of structure or structural elements, including those of the footing, piles, basement walls, etc., where strength of structural material is decisive (case B)

(*c*) failure in the ground, where strength of ground is decisive (case C).

Cases A, B, and C are cases for the ultimate limit state verification, and should not be confused with load cases, as described in Section 9.1.

Verification of the static equilibrium and strength is generally treated in *clause 9.4.1*, appropriate combinations of actions are provided in *clause 9.4.2* and relevant sets of partial factors for cases A, B and C are recommended in *clause 9.4.3*. Detailed information for verification case C, failure in the ground, is provided in *Eurocode 7, Geotechnical design*.

Clause 9.4.1
Clause 9.4.2
Clause 9.4.3

Verification of static equilibrium and strength

According to *clause 9.4.1(1)P*

Clause 9.4.1(1)P

$$E_{d,dst} \leq E_{d,stb} \qquad (9.8)$$

and when considering a limit state of static equilibrium or of the gross displacement of the structure as a rigid body, it should be verified that the design effects of destabilising actions $E_{d,dst}$ are less than the design effects of the stabilising actions $E_{d,stb}$.

Consider, for example, the simple beam structure shown in Fig. 9.1, where G_R represents the self-weight of the structure between the supports AB (the 'anchor' part), and G_S, the self-weight of the cantilever BC; assume also that the supports are unable to carry tensile force.

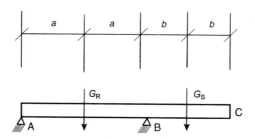

Fig. 9.1. Beam structure – static equilibrium

To satisfy expression (9.8)

$$G_{Rd}a > G_{Sd}b \qquad (D9.5)$$

where

$$G_{Rd} = \gamma_{G,inf}G_R \qquad (D9.6)$$

$$G_{Sd} = \gamma_{G,sup}G_s \qquad (D9.7)$$

According to *clause 9.4.1(2)P*

$$E_d \leq R_d \qquad (9.9)$$

Clause 9.4.1(2)P

and when considering a limit state of rupture or excessive deformation of a section, a member or connection, it should be verified that the design value of the effects of actions such as internal forces or moments E_d is less than the design value for the corresponding resistance R_d. It should be noted that for the verification of structures in the Eurocodes, $\gamma_{G,sup}$ and $\gamma_{G,inf}$ provided in *Table 9.2* are different for stability and strength checks. In the example provided (Fig. 9.1), the multiplier for the cantilever section BC will be 1·1 ($\gamma_{G,sup}$) and that for the 'anchor' span AB will be 0·9 ($\gamma_{G,inf}$).

Combination of actions

In accordance with *clause 9.4.2(1)P*, the design values of the effects of actions E_d should be determined by combining the values of actions that occur simultaneously. It is very unlikely that all loads will act at their most unfavourable values at the same time. To allow for this, the characteristic values of particular actions will need modification in accordance with Chapter 4 and further guidance provided in this chapter (*clause 9.4.2(3)*). In the expressions given below the '+' symbol does not have the normal mathematical meaning, as the directions of loads could be different. The true meaning of '+' is 'combined with', and all the expressions given in this section of the handbook may refer to actions or action effects (*clause 9.4.2(5)*).

Clause 9.4.2(1)P

Clause 9.4.2(3)

Clause 9.4.2(5)

(a) Persistent and transient situations (the fundamental combination) for ultimate limit state verifications other than those relating to fatigue are symbolically represented as follows

$$\sum_{j\geq1} \gamma_{Gj}G_{kj} \text{ '+' } \gamma_P P_k \text{ '+' } \gamma_{Q1}Q_{k1} \text{ '+' } \sum_{i>1}\gamma_{Qi}\psi_{0i}Q_{ki} \qquad (9.10)$$

This combination assumes that a number of variable actions are acting simultaneously. Q_{k1} is the dominant action and this is combined with the combination value of the accompanying actions Q_{ki}. When the dominant action is not obvious, each variable action should be considered in turn as the dominant action. See also Table 9.5 (*clause 9.4.2(2)*).

Clause 9.4.2(2)

Expression (9.10) is an amalgamation of the two separate load combinations

$$\sum_{j\geq1} \gamma_{Gj}G_{kj} \text{ '+' } \gamma_P P_k \text{ '+' } \gamma_{Q1}\psi_{01}Q_{k1} \text{ '+' } \sum_{i>1}\gamma_{Qi}\psi_{0i}Q_{ki} \qquad (9.10a)$$

$$\sum_{j\geq1} \xi_j\gamma_{Gj}G_{kj} \text{ '+' } \gamma_P P_k \text{ '+' } \gamma_{Q1}Q_{k1} \text{ '+' } \sum_{i>1}\gamma_{Qi}\psi_{0i}Q_{ki} \qquad (9.10b)$$

where ξ is a reduction factor for γ_{Gj} within the range 0·85 to 1.

The more unfavourable of *expressions (9.10a) and (9.10b)* may be applied instead of *expression (9.10)*, but only under conditions defined by the relevant National Application Documents. *Expressions (9.10a) and (9.10b)* will always give a lower design value for load effect than *expression (9.10)*. *Expression (9.10a)* will be the more unfavourable when the variable action is greater than the permanent action, while *expression (9.10b)* will be the more unfavourable when the permanent action is greater than the variable action.

Figure 9.2 demonstrates the effects of this reduction in a very simple example for the combination of one permanent load G and one variable load Q. In this example it is shown that for thicker concrete slabs *expression (9.10a)* is decisive while for thinner concrete slabs and timber beams *expression (9.10b)* is decisive. If these equations are used in the verification process, then both expressions should be used for the particular load combination and the more unfavourable one should be chosen.

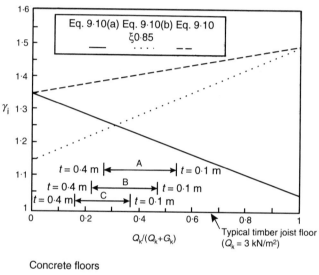

Concrete floors

Thicknesses (t) range from 0·1 m to 0·4 m for
Q_k = A 4 kN/m^2 for Categories A (balconies) and C2 (Range A)
　　　A 3 kN/m^2 for Categories B and C1 (Range B)
　　　A 2 kN/m^2 for Category A (general C)

$$\gamma_i = \frac{\text{Design load from Eq. 9·10(a) or (b)}}{(Q_k + G_k)}$$

Fig. 9.2.　*Effect of ξ factor*

(b)　The accidental design situation is symbolically represented as follows

$$\sum_{j \geq 1} \gamma_{GAj} G_{kj} \text{ '+' } \gamma_{PA} P_k \text{ '+' } A_d \text{ '+' } \psi_{11} Q_{k1} \text{ '+' } \sum_{i \geq 1} \psi_{2i} Q_{ki} \qquad (9.11)$$

This combination considers that:

- accidents are unintended events such as explosions, fire or vehicular impact, which are of short duration and have a low probability of occurrence
- a degree of damage is generally acceptable in the event of an accident
- accidents generally occur when structures are in use.

Hence, to provide a realistic accidental load combination, accidental loads are applied directly, with the frequent and quasi-permanent combination values used for the dominant and other variable actions respectively. When the dominant action is not obvious, each variable action should be considered in turn as the dominant action (*clause 9.4.2(1)*).

The combinations for accidental design situations either involve an explicit design value of accidental action A_d (e.g. impact) or refer to a situation after an accidental event ($A_d = 0$). For fire situations A_d refers to the design value of the indirect thermal action as determined by the action parts of Eurocode 1 (*clause 9.4.2(4)*).

(c)   The seismic design situation is symbolically represented as follows

$$\sum_{j \geq 1} G_{kj} \,'+'\, P_k \,'+'\, \gamma_1 A_{Ed} \,'+'\, \sum_{i \geq 1} \psi_{2i} Q_{ki} \qquad (9.12)$$

In this case the design value of the seismic action A_{Ed} is multiplied by an importance factor γ_1, and the quasi-permanent combination values for all the accompanying variable actions. Reference should be made to *Eurocode 8, Design of structures for earthquake resistance*, and its handbook.

In the case where two or more components of the load effect are partially or fully correlated, e.g. in the case of an eccentrically loaded column (see Fig. 9.3), the partial factor for actions to any favourable component could be reduced by up to 20%, if allowed by the appropriate NAD (*clause 9.4.2(6)*).

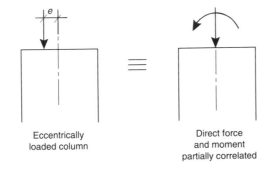

Fig. 9.3.   *Partially correlated components*

Actions arising from imposed deformations should be considered where relevant (*clause 9.4.2(7)*).

No further comment is necessary on *clause 9.4.2(8)*.

Partial factors

Table 9.2 provides the partial factors for persistent, transient and accidental design situations for the ultimate limit state verification of buildings. Table 9.1 (case A for loss of static equilibrium where the strength of the structural material or ground is insignificant), Table 9.2 (case B for failure of structure or structural elements, including piles, basement walls, etc., governed by strength of structural material) and Table 9.3 (case C for failure in the ground) reproduce the information of the key *Table 9.2*.

When using these tables, permanent actions that produce unfavourable effects (i.e. increase the effects of the variable actions) should be represented by their upper design values (using $\gamma_{G,sup} = 1\cdot1$ for static equilibrium verifications and $1\cdot35$ for strength verifications), and those that produce favourable effects (i.e. decrease the effects of the variable actions) should be represented by the lower

Table 9.1. Partial factors: ultimate limit states for buildings

Actions	Symbol	Situation/load combination	
		Persistent/transient	Accidental
Permanent actions: self-weight of structural and non-structural components, permanent actions caused by ground, groundwater and free water			
unfavourable	$\gamma_{G,sup}$	[1·10]	[1·00]
favourable	$\gamma_{G,inf}$	[0·90]	[1·00]
Variable actions			
unfavourable	γ_Q	[1·50]	[1·00]
Accidental actions	γ_A		[1·00]

Case A – loss of static equilibrium; strength of structural material or ground insignificant; see ('Verification of static equilibrium and strength').

Table 9.2. Partial factors: ultimate limit states for buildings

Actions	Symbol	Situation/load combination	
		Persistent/transient	Accidental
Permanent actions: self-weight of structural and non-structural components, permanent actions caused by ground, groundwater and free water			
unfavourable	$\gamma_{G,sup}$	[1·35]*	[1·00]
favourable	$\gamma_{G,inf}$	[1·00]	[1·00]
Variable actions			
unfavourable	γ_Q	[1·50]	[1·00]
Accidental actions	γ_A		[1·00]

Case B – failure of structure or structural elements, including those of the footing, piles, basement walls, etc., governed by strength of structural material; see ('Verification of static equilibrium and strength').

*In this verification the characteristic values of all permanent actions from one source are multiplied by [1·35] if the total resulting action effect is unfavourable, and by [1·00] if the total resulting action effect is favourable.

design values (using $\gamma_{G,inf} = 0.9$ for static equilibrium verifications and 1·0 for strength verifications) (*clause 9.4.3(1)P*). The footnote to Table 9.2 (case B) provides a simplification.

Clause 9.4.3(1)P

Where the results of verification may be very sensitive to variations of the magnitude of a permanent action from position to position in the structure, the unfavourable and the favourable parts of this action should be considered as individual actions, in particular in the verification of static equilibrium (*clause 9.4.3(2)P*).

Clause 9.4.3(2)P

The design should always be verified for each case (A, B and C) separately; for lateral earth pressure actions the design ground properties may be introduced in accordance with *Eurocode 7, Geotechnical design.*

Table 9.3. Partial factors: ultimate limit states for buildings

Actions	Symbol	Situation/load combination	
		Persistent/transient	Accidental
Permanent actions: self-weight of structural and non-structural components, permanent actions caused by ground, groundwater and free water			
unfavourable	$\gamma_{G,sup}$	[1·00]	[1·00]
favourable	$\gamma_{G,inf}$	[1·00]	[1·00]
Variable actions unfavourable	γ_Q	[1·30]	[1·00]
Accidental actions	γ_A		[1·00]

Case C — failure of the ground.

ψ factors

Table 9.4 reproduces *Table 9.3*, which gives values for buildings to be used in the load combination expressions (*clause 9.4.4*).

Clause 9.4.4

Table 9.4. ψ factors for buildings

Action	ψ_0	ψ_1	ψ_2
Imposed loads in buildings*			
Category A: domestic, residential	[0·7]	[0·5]	[0·3]
Category B: offices	[0·7]	[0·5]	[0·3]
Category C: congregation areas	[0·7]	[0·7]	[0·6]
Category D: shopping	[0·7]	[0·7]	[0·6]
Category E: storage	[1·0]	[0·9]	[0·8]
Traffic loads in buildings			
Category F: vehicle weight ≤30 kN	[0·7]	[0·7]	[0·6]
Category G: 30 kN < vehicle weight ≤160 kN	[0·7]	[0·5]	[0·3]
Category H: roofs	[0]	[0]	[0]
Snow loads on buildings	[0·6][†]	[0·2][†]	[0][†]
Wind loads on buildings	[0·6][†]	[0·5][†]	[0][†]
Temperature (non-fire) loads in buildings[‡]	[0·6][†]	[0·5][†]	[0][†]

*For combination of imposed loads in multi-storey buildings, see ENV 1991-2-1.
[†] Modification for different geographical regions may be required.
[‡] See ENV 1991-2-5.

Using *expression (9.10)* and Tables 9.2 and 9.4, the expressions may be represented as given in Table 9.5 for the example of permanent, imposed and wind loads.

Worked examples 9.1, 9.2, 9.3 at the end of this chapter explain the use of Tables 9.1, 9.2, 9.3 and 9.4 (i.e. *Tables 9.2 and 9.3*).

Table 9.5.   Partial safety factors for load combinations: ultimate limit state

Load combination	Permanent load		Variable load		
			Imposed		Wind
	Unfavourable	Favourable	Unfavourable	Favourable	
Permanent + imposed	1·35	1·0	1·50	0	—
Permanent + wind	1·35	1·0	—	—	1·50
Permanent + imposed (dominating) + wind	1·35	1·0	1·50	0	0·6 × 1·50
Permanent + wind (dominating) + imposed	1·35	1·0	0·7 × 1·50	0	1·50

Simplified verification for building structures

Eurocode 1, Part 1 permits the following simplified combinations as an alternative to the combination given in Section 9.4.2 (*clause 9.4.5*).

Clause 9.4.5

(a)   Design situation with one variable action only

$$\sum_{j\geq1} \gamma_{Gj}G_{kj} \,'+'\, [1\cdot5]Q_{k1} \tag{9.13}$$

(b)   Design situation with two or more variable actions

$$\sum_{j\geq1} \gamma_{Gj}G_{kj} \,'+'\, [1\cdot35]\sum_{i\geq1} Q_{ki} \tag{9.14}$$

For normal building structures the above expressions for ultimate limit state may also be represented as shown in Table 9.6.

Table 9.6.   Partial safety factors for simplified load combinations: ultimate limit state

Load combination	Permanent load		Variable load		
			Imposed		Wind
	Adverse	Beneficial	Adverse	Beneficial	
Permanent + imposed	1·35	1·00	1·50	0	—
Permanent + wind	1·35	1·00	—	—	1·50
Permanent + imposed + wind	1·35	1·00	1·35	0	1·35

9.5. Serviceability limit states

Verification of serviceability
According to *clause 9.5.1*

Clause 9.5.1

$$E_d \leq C_d \qquad (9.15)$$

it should be verified that the design value for action effects E_d determined on the basis of one of the combinations defined below does not exceed a nominal value or a function of certain design properties of materials C_d related to the design effects of actions considered. The explanation in Eurocode 1, Part 1 for C_d, reproduced above, may be confusing and a more correct definition covering all cases is 'serviceability constraints'.

Combination of actions
The following three combinations are normally considered for the verification of the serviceability limit states (*clauses 9.5.2(1) and 9.5.2(2)*).

Clauses 9.5.2(1) and 9.5.2(2)

(a) The rare combination symbolically represented as

$$\sum_{j\geq 1} G_{kj} \text{'+'} P_k \text{'+'} Q_{k1} \text{'+'} \sum_{i>1} \psi_{0i} Q_{ki} \qquad (9.16)$$

represents a combination of service loads that can be considered rare. This combination should be used, for example, in accordance with Eurocode 3, for the limiting values of vertical deflection.

(b) The frequent combination symbolically represented as

$$\sum_{j\geq 1} G_{kj} \text{'+'} P_k \text{'+'} \psi_{11} Q_{k1} \text{'+'} \sum_{i>1} \psi_{2i} Q_{ki} \qquad (9.17)$$

represents a combination likely to occur relatively frequently. This combination should be used, for example, in accordance with Eurocode 2 in connection with some aspects of cracking of concrete.

(c) The quasi-permanent combination symbolically represented as

$$\sum_{j\geq 1} G_{kj} \text{'+'} P_k \text{'+'} \sum_{i\geq 1} \psi_{2i} Q_{ki} \qquad (9.18)$$

Partial factors
Unless otherwise stated (e.g. in Eurocodes 2–9), the partial factors for serviceability limit states are equal to 1·0 (*clause 9.5.3*).

Clause 9.5.3

ψ factors
For serviceability limit state verifications Eurocode 1, Part 1 in *clause 9.5.4* recommends the use of Table 9.4 (*Table 9.3*).

Clause 9.5.4

Simplified verification for building structures
Eurocode 1, Part 1 permits the following simplified combinations as an alternative to the combinations given in 'Combination of actions' above (*clause 9.5.5*).

Clause 9.5.5

(a) Design situations with one variable action only

$$\sum_{j\geq 1} G_{kj} \text{'+'} Q_{k1} \qquad (9.19)$$

(b)   Design situations with two or more variable actions

$$\sum_{j \geq 1} G_{kj} \text{'+'} [0.9] \sum_{i \geq 1} Q_{ki}$$ (9.20)

There are no further comments on *clause 9.5.5(2)*.   *Clause 9.5.5(2)*

Partial factors for materials

Eurocode 1, Part 1 refers the user to Eurocodes 2 to 9, and the reader should consult these together with the appropriate handbooks in this series (*clause 9.5.6*).   *Clause 9.5.6*

Worked example 9.1—framed structure

For the frame shown in Fig. 9.4, identify the various load arrangements to check the overall stability of the structure. Assume that the building will be used as an office.

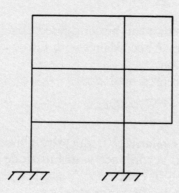

Fig. 9.4.   Frame configuration

Note that the load arrangements for the design of the individual elements could be different.

NOTATION
Characteristic loads/m

G_{kr} self-weight for the roof
G_{kf} self-weight for floors
Q_{kr} imposed load for the roof
Q_{kf} imposed floor loads

Characteristic load/frame

W_k wind loads roof or floor

LOAD CASES
The fundamental load combination that should be used is from *clause 9.4.2*   *Clause 9.4.2*

$$\sum_{j \geq 1} \gamma_{Gj} G_{kj} \text{'+'} \gamma_P P_k \text{'+'} \gamma_{Q1} Q_{k1} \text{'+'} \sum_{i > 1} \gamma_{Qi} \psi_{0i} Q_{ki}$$ (9.10)

As the stability of the structure will be sensitive to a possible variation of self-weights, it will be necessary to allow for this in accordance with Table 9.2 (also *Table 9.2*). Thus

$$\gamma_{G,inf} = 1\cdot0$$
$$\gamma_{G,sup} = 1\cdot35$$
$$\gamma_Q = 1\cdot5$$

$\psi_0 = 0\cdot7$ for imposed loads (offices) from Table 9.4 (i.e. *Table* 9.3)

$\psi_0 = 0\cdot6$ for snow and wind loads for buildings from Table 9.4 (i.e. *Table* 9.3)

LOAD CASE 1
Treat the wind load as the dominant action (see Fig. 9.5).

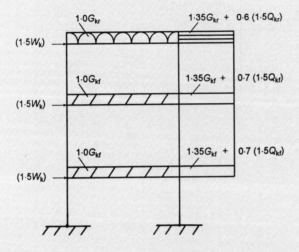

Fig. 9.5. Load case 1

LOAD CASE 2
Treat the imposed load on the roof as the dominant load (see Fig. 9.6).

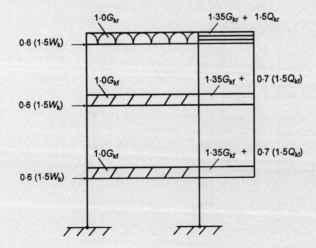

Fig. 9.6. Load case 2

LOAD CASE 3

Treat the imposed load on the floors as the dominant load (see Fig. 9.7).

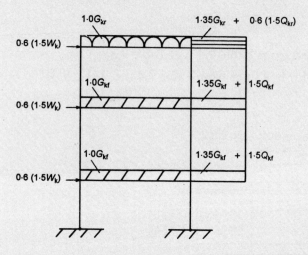

Fig. 9.7.   *Load case 3*

LOAD CASE 4

Consider the case without wind loading, treating the imposed floor loads as the primary load (see Fig. 9.8).

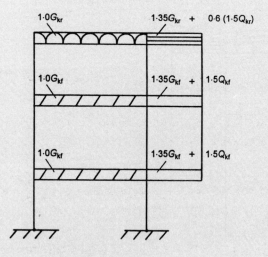

Fig. 9.8.   *Load case 4*

LOAD CASE 5
Consider the case without wind loading, treating the imposed roof load as the primary load (see Fig. 9.9).

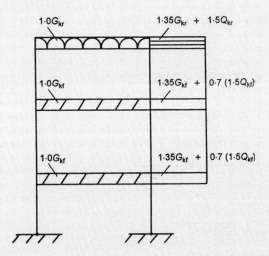

Fig. 9.9.   Load case 5

Note: When the wind loading is reversed, another set of arrangements will need to be considered.

Worked example 9.2—continuous beam
Identify the various load arrangements for the ultimate limit state for the design of the four-span continuous beam shown in Fig. 9.10. Assume that spans 1–2 and 2–3 are for domestic and residential use and spans 3–4 and 4–5 for office use.

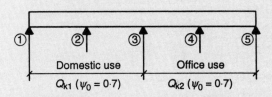

Fig. 9.10.   Beam configuration

NOTATION
G_k characteristic self-weight/m
Q_{k1} characteristic imposed load/m (domestic and residential use)
Q_{k2} characteristic imposed load/m (office use)

LOAD CASES
The fundamental load combination that should be used is from *clause 9.4.2*

$$\sum_{j\geq1} \gamma_{Gj}G_{kj} \text{`+'} \gamma_P P_k \text{`+'} \gamma_{Q1}Q_{k1} \text{`+'} \sum_{i>1} \gamma_{Qi}\psi_{0i}Q_{ki} \qquad (9.10)$$

Clause 9.4.2

From the footnote to Table 9.2 (i.e. footnote 3 of *Table 9.2*) the same value of self-weight may be applied to all spans, i.e. $1.35G_k$.

The load cases that should be considered for the imposed loads are

(*a*)   alternative spans loaded
(*b*)   adjacent spans loaded.

The various load arrangements are shown in Fig. 9.11.

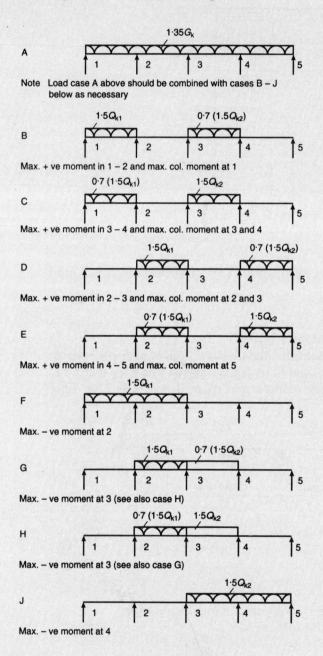

Fig. 9.11.   Load cases

Worked example 9.3—concrete basement wall

A concrete basement wall as shown in Fig. 9.12 is subjected to the earth pressure originating from a frictional (non-cohesive) material with the characteristic angle of friction $\varphi_k = 35°$, the characteristic unit weight $\gamma_k = 18$ kN/m^3 and a characteristic surcharge $p_k = 10$ kN/m^2. The objective is to determine the design moment in the wall shown in Fig. 9.12.

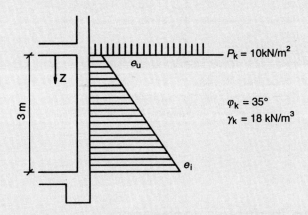

Fig. 9.12. *Concrete basement wall*

CASE A, TABLE 9.1 (i.e. case A, *Table 9.2*)
Case A is irrelevant to the present problem.

CASE B, TABLE 9.2 (i.e. case B, *Table 9.2*)
For the angle of internal friction $\varphi_d = \varphi_k = 35°$ the relevant earth pressure coefficients are $K_p = 0.218$ and $K\gamma = 0.211$. The characteristic earth pressure is shown diagramatically in Fig. 9.12 and given mathematically in Table 9.5.

Table 9.5. Earth pressure for case B

	Characteristic	Design
e_u: kN/m^2	$0.218 \times 10 = 2.18$	$2.18 \times 1.5 = 3.27$
e_i: kN/m^2	$0.218 \times 10 \times 1.5 + 3 \times 18 \times$ $0.211 \times 1.35 = 18.65$	$13.57 \times 1.35 = 18.31$

This case needs to be considered, as the earth pressure is unfavourable to the structure. Consequently, the design values of the earth pressures are determined by multiplying by partial factors in accordance with Table 9.2 (see also *Table 9.2*). The maximum moment on the wall is accordingly determined with $e_u = 3.27$ kN/m^2 and $e_i = 18.65$ kN/m^2 (see Fig. 9.12).

CASE C, TABLE 9.3 (i.e. case C, *Table 9.2*)
For a characteristic angle of the internal friction $\varphi_k = 35°$, the design friction angle is $\varphi_d = \arctan(\tan 35°/1.25) = 29° \times 3$. The relevant earth pressure coefficients are $K_p = 0.282$ and $K_\gamma = 0.275$. The design value of the surcharge is $P_d = 10 \times 1.3 = 13$ kN/m^2. The design earth pressures are given in Table 9.6.

Table 9.6. Earth pressure for case C

	Characteristic	Design
e_u: kN/m^2	$0.282 \times 10 = 2.82$	$2.82 \times 1.3 = 3.67$
e_i: kN/m^2	$0.282 \times 10 + 0.275 \times 3 \times 18$	$2.82 \times 1.3 + 14.05 \times 1 = 18.52$

The maximum moment in the wall is accordingly determined with $e_u = 3.67$ kN/m^2 and $e_i = 18.52$ kN/m^2 (see Fig. 9.12).

The determining case is determined after examining the moments in the wall due to cases B and C (i.e. Tables 9.5 and 9.6 in this example).

Appendix 1: Inconsistencies in design resistance formulae between Eurocode 1, Part 1 and Eurocodes 2, 3 and 4

Clause 9.3.5 states the verification formulae for design resistance factors as follows

$$R_d = R(X_k/\gamma_M, a_{nom}) \qquad \text{for Eurocode 2} \qquad (9.7a)$$

$$R_d = R(X_k, a_{nom})/\gamma_R \qquad \text{for Eurocode 3} \qquad (9.7b)$$

$$R_d = R(X_k/\gamma_m, a_{nom})/\gamma_{rd} \qquad \text{for Eurocode 4} \qquad (9.7c)$$

However, according to Eurocodes 2 to 4, and the handbook (in this series) to Eurocode 4,[48] the expressions should be

$$R_d = R(X_k/\gamma_M, a_{nom}) \qquad \text{for Eurocode 2} \qquad (A.12)$$

$$R_d = R(X_k, a_{nom})/\gamma_R \qquad \text{for Eurocode 3} \qquad (A.13)$$

$$R_d = R(X_k/\gamma_m, a_{nom})/\gamma_{rd} \qquad \text{for Eurocode 4} \qquad (A.14)$$

The apparent differences may be explained as follows:

(a) γ_M is the same as γ_R
(b) with regard to the equations for Eurocode 4 (i.e. A.14) and (9.7c) above, the definitions for γ_{rd} and γ_{Rd} in Eurocode 1, Part 1 are as follows

- γ_{rd} is the partial factor associated with the uncertainty of the resistance model and the dimensional variations
- γ_{Rd} is the partial factor associated with the uncertainty of the resistance model.

Hence by examining the two expressions (A.14 and 9.7c), it is seen that the 'correct scientific' term to use is γ_{rd}.

APPENDIX A

The Construction Products Directive (89/106/EEC)

As compliance with Structural Eurocodes will satisfy the requirements of the Construction Products Directive in respect of mechanical resistance, this Appendix provides a short introduction to the Construction Products Directive (Directive 89/106/EEC),[49] the essential requirements contained in the Directive and the Interpretative Document to the first essential requirement, 'Mechanical resistance and stability'.

The following abbreviations are used in this Appendix

CEC	Commission of European Communities
CEN	European Committee for Standardisation
CPD	Construction Products Directive
EFTA	European Free Trade Association
EN	European Standard
ENV	European Pre-Standard
EU	European Union
PT	Project Team
SC	Sub-committee
TC	Technical Committee.

A.1. Compliance of the Structural Eurocodes with the Construction Products Directive

Compliance with the Structural Eurocodes will satisfy the requirement of the Construction Products Directive in respect of 'Mechanical resistance and stability'.

A.2. Purpose of the Construction Products Directive

A major objective of the Commission of European Communities (CEC) is the removal of barriers to trade across national borders and implementation of the European Internal Market. This Market is characterised by the free movement of goods, services, capital and people. To realise this objective, approximation of laws, regulations and administrative provisions for European Union (EU) member states is necessary. The Construction Products Directive (CPD), relating to construction products, which was adopted by the CEC on 21 December 1988, is an important element of this approximation process. The CPD has been transferred into national law by the majority of EU member states.

The main objective of the CPD is to provide a legal basis for the attestation of conformity (see Fig. A.1) so that free movement of the relevant product is ensured. The scope of the CPD is construction products, which are defined as products that are produced for incorporation in a permanent manner in construction works, including both building and civil engineering works, insofar as the essential requirements relate to them.

Position of Category A — Standards for design and execution in the European system for the approximation of laws, regulations and administrative provisions relating to construction products

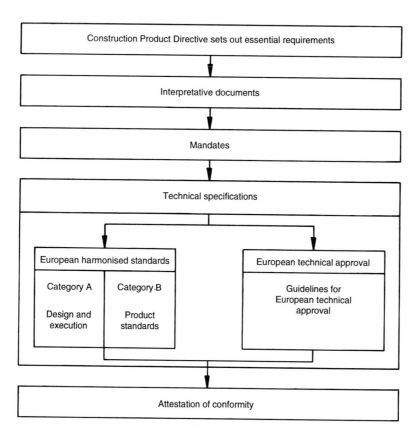

Fig. A.1. *The European system for the approximation of laws, regulations and administrative provisions relating to construction products*

A.3. The essential requirements

The essential requirements[49] apply to construction works, not to construction products as such, but they will influence the technical characteristics of these products.

Thus, construction products must be suitable for construction works which, as a whole and in their separate parts, are fit for their intended use, account being taken of economy, and which satisfy the essential requirements where the works are subject to regulations containing such requirements.

The essential requirements relate to

- mechanical resistance and stability
- safety in case of fire
- hygiene, health and environment
- safety in use
- protection against noise
- energy economy and heat retention.

These requirements must, subject to normal maintenance, be satisfied for an economically reasonable working life.

The essential requirements may give rise to the establishment of classes of a construction product corresponding to different performance levels, to take account of possible differences in geographical or climatic conditions or in ways of life as well as different levels of protection that may prevail at national, regional or local level. Member states may require performance levels to be observed in their territory only within the classification thus established.

The Structural Eurocodes relate to the essential requirement for 'Mechanical resistance and stability', which states:

'The construction works must be designed and built in such a way that the loadings that are liable to act on it during its construction and use will not lead to any of the following:

(a) collapse of the whole or part of the work
(b) major deformations to an inadmissible degree
(c) damage to other parts of the works or to fittings or installed equipment as a result of major deformation of the load-bearing construction
(d) damage by an event to an extent disproportionate to the original cause.'

The essential requirements are given concrete (i.e. quantitative) form in Interpretative Documents (IDs), which will create the necessary links between those requirements and, for example, the mandates to draw up European Standards for particular construction products (Fig. A.1). IDs for each of the six essential requirements are available. ID 1 refers to 'Mechanical resistance and stability'.

A.4. Methods of satisfying the essential requirements

In practice a product is fit for the intended use when it permits the works in which it is incorporated to satisfy the applicable essential requirements; a product is presumed to be fit for its intended use if it bears the European Community marking (see Fig. A.2)

Fig. A.2. *The EC mark signifying a product complies with the relevant standard or technical approval*

which declares the conformity of the product to technical specifications (see Fig. A.1). These specifications comprise:

- harmonized Standards (Article 7 of the CPD) established by the European Committee for Standardization (CEN) on the basis of mandates
- the European Technical Approval (Article 8 of the CPD).

Article 4 of the CPD allows referencing to National Standards, but only where harmonized European Specifications do not exist.

A.5. Interpretative Document 1: Mechanical resistance and stability[50]

With regard to the design of structures, ID 1 provides

- a definition of the general terms used in all the IDs
- a precise definition of the essential requirement, 'Mechanical resistance and stability'
- the basis for the verification of satisfaction of the requirement 'Mechanical resistance and stability'.

With regard to the verification along with other rules, ID 1 states in clause 3.1(3):

'The satisfaction of the essential requirements is assured by a number of interrelated measures concerned in particular with

- the planning and design of the works, the execution of the works and necessary maintenance
- the properties, performance and use of the construction products.'

Thus, ID 1 distinguishes between the design of the buildings and civil engineering works as a whole and the technical properties of the construction products themselves. In clause 4 of ID 1, design and product standards are defined as follows.

- '*Category A*: These are standards,which concern the design and execution of buildings and civil engineering works and their parts, or particular aspects thereof, with a view to the fulfilment of the essential requirements as set out in the Council Directive 89/106/EEC. Category A standards should be taken into consideration within the scope of the Directive as far as differences in laws, regulations and administrative provisions of Member States prevent the development of harmonised product standards.
- *Category B*: These are technical specifications and guidelines for European technical approval which exclusively concern construction products subject to an attestation of conformity and marking according to Articles 13, 14 and 15 of the Council Directive 89/106/EEC. They concern requirements with regard to performance and/or other properties including durability of those characteristics that may influence the fulfilment of the essential requirements, testing, and compliance criteria of a product. Category B standards that concern a family of products, or several families of products are of different character and are called horizontal (Category B) standards.'

ID 1 further states that

'the assumptions made in Category A standards on the one side and those made in Category B specifications on the other side shall be compatible to each other'.

A.6. Category A standards

The Structural Eurocodes are considered as one group of Category A standards.

APPENDIX B

The proposed complete Eurocode suite

The proposed Eurocode suite, all listed as ENVs, will eventually be transposed to EN status. 'Eurocode programme' in the Introduction of this handbook gives approximate dates for the transposition.

ENV 1991	*Eurocode 1, Basis of design and actions on structures*
ENV 1991-1	Part 1, Basis of design
ENV 1991-2-1	Part 2-1, Actions on structures, Densities, self-weight and imposed loads
ENV 1991-2-2	Part 2-2, Actions on structures, Actions on structures exposed to fire
ENV 1991-2-3	Part 2-3, Actions on structures, Snow loads
ENV 1991-2-4	Part 2-4, Actions on structures, Wind actions
ENV 1991-2-5	Part 2-5, Actions on structures, Thermal actions
ENV 1991-2-6	Part 2-6, Actions on structures, Construction loads and deformations imposed during construction
ENV 1991-2-7	Part 2-7, Actions on structures, Accidental actions
ENV 1991-3	Part 3, Traffic loads on bridges
ENV 1991-4	Part 4, Actions in silos and tanks
ENV 1991-5	Part 5, Actions induced by cranes and machinery

ENV 1992	*Eurocode 2, Design of concrete structures*
ENV 1992-1-1	Part 1-1, General rules, General rules and rules for buildings
ENV 1992-1-2	Part 1-2, General rules, Structural fire design
ENV 1992-1-3	Part 1-3, General rules, Precast concrete elements and structures
ENV 1992-1-4	Part 1-4, General rules, Structural lightweight aggregate concrete
ENV 1992-1-5	Part 1-5, General rules, Unbonded and external tendons in buildings
ENV 1992-1-6	Part 1-6, General rules, Plain concrete structures
ENV 1992-2	Part 2, Reinforced and prestressed concrete bridges
ENV 1992-3	Part 3, Concrete foundations
ENV 1992-4	Part 4, Liquid-retaining and containment structures
ENV 1992-5	Part 5, Marine and maritime structures
ENV 1992-6	Part 6, Massive structures

ENV 1993	Eurocode 3, Design of steel structures
ENV 1993-1-1	Part 1-1, General rules, General rules and rules for buildings
ENV 1993-1-1/A1	Part 1-1/A1, General rules, General rules and rules for buildings Annexes D and K-revised
ENV 1993-1-1/A2	Part 1-1/A2, General rules, General rules and rules for buildings Annexes G, H, J-revised, N and Z
ENV 1993-1-2	Part 1-2, General rules, Structural fire design
ENV 1993-1-3	Part 1-3, General rules, Cold formed thin gauge members and sheeting
ENV 1993-1-4	Part 1-4, General rules, The use of stainless steels
ENV 1993-2	Part 2, Bridges and plated structures
ENV 1993-3	Part 3, Towers, masts and chimneys
ENV 1993-4	Part 4, Tanks, silos and pipelines
ENV 1993-5	Part 5, Piling
ENV 1993-6	Part 6, Crane structures
ENV 1993-7	Part 7, Marine and maritime structures
ENV 1993-8	Part 8, Agricultural structures

ENV 1994	Eurocode 4, Design of composite steel and concrete structures
ENV 1994-1-1	Part 1-1, General rules, General rules and rules for buildings
ENV 1994-1-2	Part 1-2, General rules, Structural fire design
ENV 1994-2	Part 2, Bridges

ENV 1995	Eurocode 5, Design of timber structures
ENV 1995-1-1	Part 1-1, General rules, General rules and rules for buildings
ENV 1995-1-2	Part 1-2, General rules, Structural fire design
ENV 1995-2	Part 2, Bridges

ENV 1996	Eurocode 6, Design of masonry structures
ENV 1996-1-1	Part 1-1, General rules, Rules for reinforced and unreinforced masonry
ENV 1996-1-2	Part 1-2, General rules, Structural fire design
ENV 1996-1-3	Part 1-3, General rules, Detailed rules on lateral loading
ENV 1996-1-4	Part 1-4, General rules, Complex shapes
ENV 1996-2	Part 2, Special design aspects
ENV 1996-3	Part 3, Simplified and simple rules
ENV 1996-4	Part 4, Short life structures

ENV 1997	*Eurocode 7, Geotechnical design*
ENV 1997-1	Part 1, General rules
ENV 1997-2	Part 2, Geotechnical design assisted by laboratory testing
ENV 1997-3	Part 3, Geotechnical design assisted by field testing
ENV 1997-4	Part 4, Specific geotechnical structures

ENV 1998	*Eurocode 8, Design provisions for earthquake resistance of structures*
ENV 1998-1-1	Part 1-1, General rules, Seismic actions and general requirements for structures
ENV 1998-1-2	Part 1-2, General rules, General rules for buildings
ENV 1998-1-3	Part 1-3, General rules, specific rules for various materials and elements
ENV 1998-1-4	Part 1-4, General rules, strengthening and repair
ENV 1998-2	Part 2, Bridges
ENV 1998-3	Part 3, Towers, masts and chimneys
ENV 1998-4	Part 4, Silos, tanks and pipelines
ENV 1998-5	Part 5, Foundations, retaining structures and geotechnical aspects

ENV 1999	*Eurocode 9, Design of Aluminium Alloy Structures*
ENV 1999-1-1	Part 1-1, General rules, General rules and rules for buildings
ENV 1999-1-2	Part 1-2, General rules, Structural fire design
ENV 1999-2	Part 2, Rules for structures susceptible to fatigue

APPENDIX C

Basic statistical terms and techniques

C.1. General

This Appendix is intended primarily for the use of those advising on the drafting of codes of practice for the construction industry (*clause 17* of *Foreword*); and for the guidance for those involved in the design of complex structures not fully covered by existing civil engineering codes of practice (*clause 19* of *Foreword*).

Clause 17

Clause 19

It is assumed that the reader of this Appendix has some knowledge of the application of statistics and probability theory in solving civil engineering problems.

The terms and concepts described are those occurring often in *Eurocode 1, Part 1, Basis of design* (in particular in *Sections 4, 5, 6 and 8* and *Annexes A and D*) and in *International Standard ISO 2394, General principles on reliability for structures*.[10] Definitions presented in this Appendix correspond to the probabilistic and general statistical terms given in International Standards,[51-59] in particular ISO 3534/1,2[55,56] and ISO 12491[59] where a more comprehensive description and other terms and definitions may be found. In addition, for the normal (Gaussian) distribution, statistical techniques used in civil engineering[59] for the estimation and tests of basic statistical characteristics are given. Also for the general three-parameter log-normal distribution,[41,47] basic statistical techniques for the estimation of fractiles (characteristic and design values) are reviewed.

C.2. Populations and samples

Basic terms and concepts

Actions, mechanical properties and dimensions are described by random variables. A random variable X (e.g. for concrete strength) may take each of a specified set of values with a known or estimated probability. As a rule, only a limited number of observations constituting a random sample $x_1, x_2, x_3, \ldots, x_n$ of size n taken from a population is available for a variable X. Population is a general statistical term used for the totality of units under consideration, e.g. for all concrete produced under specified conditions within a certain period of time. The aim of statistical methods is to make decisions concerning the properties of a population using the information derived from one or more random samples.

Sample characteristics

A sample characteristic is a quantity used to describe the basic properties of a sample. The three basic sample characteristics most commonly used in practical applications are:

- the mean m representing the basic measure of central tendency
- the variance s^2 as the basic measure of dispersion
- the coefficient of skewness a giving the degree of asymmetry.

The sample mean m is defined as the sum

$$m = (\sum x_i)/n \qquad (C.1)$$

with the summation being extended over all the n values of x_i.

Following the terminology of International Standard ISO 3534-1,[55] the *sample variance* s^2 is defined as

$$s^2 = (\sum(x_i - m)^2)/(n - 1) \qquad (C.2)$$

the summation again being extended over all the values x_i. Sample standard deviation s is the positive square root of the variance s^2. Strictly speaking, the term 'sample variance' should be used for the central moment of order 2. The reason for using the variance s^2, defined by expession (C.2), as the sample characteristic is that it is an unbiased estimator of the population variance.

The *sample coefficient of skewness a*, characterising asymmetry of the distribution, is defined as

$$a = [n(\sum(x_i - m)^3)/(n - 1)/(n - 2)]/s^3 \qquad (C.3)$$

Thus the coefficient of skewness is derived from the central moment of order 3 divided by s^3. If the sample has more distant values to the right from the mean than to the left, the distribution is said to be skewed to the right or to have positive skewness. If the reverse is true it is said to be skewed to the left or to have negative skewness.

Another important characteristic describing the dispersion of a sample is the *coefficient of variation v*, defined as the ratio of standard deviation s to the mean m

$$v = s/m \qquad (C.4)$$

However, the coefficient of variation can be effectively used only if the mean m is significantly greater than the standard deviation s. When the mean is close to zero the standard deviation, rather than the coefficient of variation, should be used as a measure of the dispersion.

Population parameters

Probability distribution is a term generally used for any function giving the probability that a variable X belongs to a given set of values.

The basic theoretical models used to describe the probability distribution of a random variable may be obtained from a random sample by increasing the sample size or by smoothing either the frequency distribution or the cumulative frequency polygon.

An idealisation of a cumulative frequency polygon is the distribution function $F(x)$ giving, for each value x, the probability that the variable X is less than or equal to x

$$F(x) = P(X \leq x) \qquad (C.5)$$

A *probability density function* f(x) is an idealisation of a relative frequency distribution. It is formally defined as the derivative (when it exists) of the distribution function

$$f(x) = dF(x)/dx \qquad (C.6)$$

The population parameters are quantities used in describing the distribution of a random variable, as estimated for one or more samples. As in the case of random samples, three basic population parameters are commonly used in practical applications:

- the mean μ representing the basic measure of central tendency
- the variance σ^2 as the basic measure of dispersion
- the coefficient of skewness α giving the degree of asymmetry.

The *population mean* μ for a continuous variable X having the probability density f(x) is defined as

$$\mu = \int x \, f(x) \, dx \qquad (C.7)$$

the integral being extended over the interval of variation of the variable X. The *population variance*, σ^2 for a continuous variable X having the probability density function f(x) is the mean of the squared deviation of the variable from its mean

$$\sigma^2 = \int (x - \mu)^2 \, f(x) \, dx \qquad (C.8)$$

The population standard deviation σ is the positive square root of the population variance σ^2.

The *population coefficient of skewness* α characterising asymmetry of the distribution is defined as

$$\alpha = \int (x - \mu)^3 \, f(x) \, dx / \sigma^3 \qquad (C.9)$$

Another important parameter is the *coefficient of variation* V of the population

$$V = \sigma/\mu \qquad (C.10)$$

The same restriction on the practical use of V applies as in the case of samples.

In the Eurocodes, a very important population parameter is the fractile x_p. If X is a continuous variable and p is a real number between 0 and 1, the *p-fractile x_p* is the value of the variable X for which the probability that the variable X is less than or equal to x_p is equal to p, and hence for which the distribution function F(x_p) is equal to p

$$P(X \le x_p) = F(x_p) = p \qquad (C.11)$$

In civil engineering the probabilities $p = 0.001$, 0.01, 0.05 and 0.10 are often used. p is often written as a percentage ($p = 0.1\%$; 1%; 5%; 10%); if this is done then p is called a percentile, and x_p is called, for example, the 5th percentile, meaning $p = 5\%$. If $p = 50\%$ then x_p is called the median.

C.3. Normal and log-normal distribution

Normal distribution
The very common normal distribution (Gaussian distribution) may often be used to approximate symmetrical bell-shaped distributions.

The normal probability density function of a continuous random variable X having the mean μ and standard deviation σ is defined on the infinite interval $\langle -\infty, +\infty \rangle$ as

$$f(x) = \exp[-(x-\mu)^2/(2\sigma^2)]/[\sigma(2\pi)^{1/2}] \qquad (C.12)$$

Civil and structural engineers should rarely, if ever, need to use expression (C.12) since tables of $f(x)$ and its cumulative distribution $F(x)$ can be found in statistical tables.

If a significant asymmetry is present however, then use of some other type of distribution, reflecting this asymmetry, should be considered. Often the three-parameter log-normal distribution is used under such circumstances. The log-normal distribution, defined on a semi-infinite interval, is generally described by three parameters: besides the mean μ and variance σ^2 a third characteristic, namely the lower or upper limit value x_0 or, alternatively, the coefficient of skewness α, may be used. In many practical applications the coefficient of skewness α can be expected to be within the interval $\langle -1, +1 \rangle$. If the limit value x_0 is known, then the distribution of a variable X may be easily transformed to normal distribution of a variable Y

$$Y = \log | X - x_0 | \qquad (C.13)$$

which can then be analysed by using tables of the normal distribution.

In civil engineering, a log-normal distribution with x_0 as the lower limit (and which consequently has a positive skewness) is often used. Moreover, it is often assumed that $x_0 = 0$ and then only two parameters (μ and σ) are involved. In this case the normal variable Y is given as

$$Y = \log X \qquad (C.14)$$

and the original variable X is assumed to have positive skewness. The coefficient of skewness is dependent on the coefficient of variation $V = \sigma/\mu$ of the variable X and is given as

$$\alpha = V^3 + 3V \qquad (C.15)$$

A log-normal distribution is close to being a normal distribution if the coefficient of skewness α is zero and the absolute value of the limit value x_0 is very large.

Standardised variable

To simplify calculation procedures, standardised variables, with means of zero and variances of 1, can frequently be used.

If the variable X has the mean μ and standard deviation σ, the corresponding standardised variable U is defined as

$$U = (X - \mu)/\sigma \qquad (C.16)$$

The distribution of the standardised variable U is called a standardised distribution; it is these that are usually tabulated. For example, the standardised normal distribution has the probability density function $\varphi(u)$ in the form

$$\varphi(u) = \exp[(-u^2)/2]/(2\pi)^{1/2} \qquad (C.17)$$

Detailed tables are available for the standardised probability distribution function $\Phi(u)$ and the standardised probability density function $\varphi(u)$ (see for example ISO 12491[59]). When one is using these tables, the p-fractile u_p of the standardised variable u is given by

$$p = \Phi(u_p) \tag{C.18}$$

It follows from expression (C.12) that the p-fractile x_p of the original variable x may be determined from the fractile u_p of the standardised variable u as

$$x_p = \mu + u_p\sigma \tag{C.19}$$

For a three-parameter log-normal distribution, selected values of fractiles u_p for commonly used probabilities p and for three coefficients of skewness $\alpha = -1$, 0 and $+1$, are given in Table C.1.

Table C.1. Selected values of the standardised variable u_p for three-parameter log-normal distribution

$\Phi(u_p)$		10^{-6}	10^{-5}	10^{-4}	10^{-3}	0·01	0·05	0·10
	$\alpha = -1$	−10·05	−8·18	−6·40	−4·70	−3·03	−1·85	−1·32
u_p	$\alpha = 0$	−4·75	−4·26	−3·72	−3·09	−2·33	−1·64	−1·28
	$\alpha = +1$	−2·44	−2·33	−2·19	−1·99	−1·68	−1·34	−1·13

Normality tests

Any assumption of a normal distribution for a variable X needs to be checked. This can be done by using various normality tests: the random sample is compared with the normal distribution having the same mean and standard deviation, and the differences between the two are tested to see whether or not they are significant. If the deviations are insignificant, then the assumption of a normal distribution is not rejected; otherwise it is rejected. There will always be a difference between the two, but whether or not this difference is significant will depend on the use to which the model will be put. It is usual to express the overall fit as a significance level, which can be thought of as the probability that the actual population does not follow the assumed normal distribution.

Various normality tests, well established in the International Standard ISO/DIS 5479,[58] may be used. The recommended significance level to be used in building is 0·05 or 0·01.

C.4. Statistical methods

Basic statistical methods used in civil engineering

As indicated above, the aim of statistical methods is to make decisions concerning properties of populations using the information derived from one or more random samples.

Basic statistical methods used in civil engineering consist of estimation techniques, methods of testing of statistical hypotheses and methods of sampling inspection. In the following, only the main estimation techniques and tests are described. More extensive description of statistical methods as used in civil engineering are given in the International Standard ISO/DIS 12491.[59]

Principles of estimation and tests

Two types of technique for estimating population parameters are generally used in construction:

(a) point estimation
(b) interval estimation.

A *point estimate* of a population parameter is given by a single number, which is the value of an estimator derived from a given sample. The best point estimate of a population parameter is unbiased (the mean of the estimator is equal to the corresponding population parameter) and efficient (variance of the unbiased estimator is a minimum).

An *interval estimate* of a population parameter is given by two numbers, and contains the parameter with a certain probability γ, called the *confidence level*. The values $\gamma = 0.90$, 0.95 or 0.99, in some cases also $\gamma = 0.75$, are recommended to be used for the quality control of building, depending on the type of variable and the possible consequences of exceeding the estimated values. Interval estimates indicate the precision of an estimate and are therefore preferable to point estimates.

Methods for testing a hypothesis concerning population parameters that are commonly used in civil engineering may be divided into two groups:

(a) comparison of sample characteristics with corresponding population parameters
(b) comparison of characteristics of two samples.

A test of a statistical hypothesis is a procedure intended to indicate whether or not a hypothesis about the distribution of one or more populations should be rejected. If results derived from random samples do not differ markedly from those expected under the assumption that the hypothesis is true, then the observed difference is said to be insignificant and the hypothesis is not rejected; otherwise the hypothesis is rejected. The recommended significance levels $\alpha = 0.01$ or 0.05 guarantee that the risk of rejection of a true hypothesis is of the same order as the risk of non-rejection of a false hypothesis.

Methods for estimating and testing of means and variances are generally covered by International Standard ISO 12491[59] and also by other ISO documents.[51-58] A normal or log-normal distribution is assumed in most of these documents. In the following, the basic methods for estimation of means and variances are summarized. Also, important statistical techniques for estimating fractiles are briefly examined in Section C.5.

Estimation of the mean

The best point estimate of the population mean μ is the sample mean m. Any interval estimate of the mean μ depends on whether or not the population standard deviation σ is known. If the population standard deviation σ is known, then the two-sided interval estimate at the confidence level $\gamma = 2p - 1$ is

$$m - u_p\sigma/n^{1/2} \leq \mu \leq m + u_p\sigma/n^{1/2} \qquad (C.20)$$

where u_p is the fractile of the normal distribution corresponding to the probability p (close to 1) given in Table 1 of ISO 12491[59] (see also ISO 2854[52]).

If the population standard deviation σ is unknown, then the two-sided interval estimate at the confidence level $\gamma = 2p - 1$ is

$$m - t_ps/n^{1/2} \leq \mu \leq m + t_ps/n^{1/2} \qquad (C.21)$$

where s is the sample standard deviation and t_p is the fractile of the t-distribution for $v = n - 1$ degrees of freedom and the probability p (close to 1), as given in Table 3 of ISO 12491[59] (see also ISO 2854[52]). In each of the above cases the one-sided interval estimate at the confidence level $\gamma = p$ may be used if only the lower or only the upper limit of the above estimates is being considered. The values of p and corresponding fractiles u_p and t_p should be specified in accordance with the chosen confidence level $\gamma = p$.

Estimation of the variance

The best point estimate of the population variance σ^2 is the sample variance s^2 (see the comment on expression (C2)). A two-sided interval estimate for the variance σ^2 at the confidence level $\gamma = p_2 - p_1$, where p_1 and p_2 are chosen probabilities, is given as

$$(n - 1)s^2/\chi_{p2}^2 \le \sigma^2 \le (n - 1)s^2/\chi_{p1}^2 \qquad (C.22)$$

where χ_{p1}^2 and χ_{p2}^2 are fractiles of the χ^2 distribution for $v = (n - 1)$ degrees of freedom corresponding to the probabilities p_1 (close to 0) and p_2 (close to 1) given in Table 2 of ISO 12941[59] (see also ISO 2854[52]). Often the lower limit of the above interval estimate for the variance σ^2 is considered to be 0 and then the confidence level γ of the estimate is $1 - p_1$. The estimate for the standard deviation σ is the square root of the variance σ^2.

Comparison of means

To test the significance of the difference between the sample mean m and a supposed population mean μ if the population standard deviation σ is known, the test value

$$u_0 = | m - \mu | n^{1/2}/\sigma \qquad (C.23)$$

is compared with the critical value u_p (Table 1 in ISO 12491[59]), which is the p-fractile of the normal distribution corresponding to the significance level $\alpha = 1 - p$ (where α is a value close to 0). If $u_0 \le u_p$, then the hypothesis that the sample has been taken from the population with the mean μ is not rejected; otherwise it is rejected.

If the population standard deviation σ is unknown, then the test value

$$t_0 = | m - \mu | n^{1/2}/s \qquad (C.24)$$

is compared with the critical value t_p (Table 3 in ISO 12491[59]), which is the p-fractile of the t-distribution for the $v = n - 1$ degrees of freedom corresponding to the significance level $\alpha = 1 - p$ (where α is a small value close to 0). If $t_0 \le t_p$, then the hypothesis that the sample has been taken from the population with the mean μ is not rejected; otherwise it is rejected.

To test the difference between the means m_1 and m_2 of two samples of size n_1 and n_2 respectively, which have been taken from two populations having the same population standard deviation σ, the test value

$$u_0 = | m_1 - m_2 | (n_1 n_2)^{1/2}/[\sigma(n_1 + n_2)^{1/2}] \qquad (C.25)$$

is compared with the critical value u_p (Table 1 in ISO 12491[59]), which is the p fractile of the normal distribution corresponding to the significance level $\alpha = 1 - p$ (α is a value close to 0). If $u_0 \le u_p$ then the hypothesis that both samples have been taken from populations with the same (although unknown) mean μ is not rejected; otherwise it is rejected.

If the standard deviation σ of both populations is the same, but unknown, then it is necessary to use the sample standard deviations s_1 and s_2. The test value

$$t_0 = |m_1 - m_2| \, [(n_1 + n_2 - 2)(n_1 n_2)]^{1/2} / \{[(n_1 - 1)s_1^2 + (n_2 - 1)s_2^2](n_1 + n_2)\}^{1/2} \tag{C.26}$$

is compared with the critical value t_p (Table 3 in ISO 12491[59]), which is the fractile of the t-distribution for the $v = n_1 + n_2 - 2$ degrees of freedom corresponding to the significance level $\alpha = 1 - p$ (a small value close to 0). If $u_0 \leq u_p$, then the hypothesis that the samples have been taken from the populations with the same (though unknown) mean μ is not rejected; otherwise it is rejected.

For two samples of the same size $n_1 = n_2 = n$, for which observed values may be meaningfully coupled (paired observations), the difference between the sample means may be tested using the differences of coupled values $w_i = x_{1i} - x_{2i}$. The mean m_w and standard deviation s_w are first found and then the test value

$$t_0 = |m_w| \, n^{1/2} / s_w \tag{C.27}$$

is compared with the critical value t (Table 3 in ISO 12491[59]), which is the fractile of the t-distribution for the $v = n - 1$ degrees of freedom corresponding to the significance level $\alpha = 1 - p$ (α is a small value close to 0). If $t_0 \leq t_p$, then the hypothesis that both samples are taken from populations with the same (unknown) mean μ is not rejected; otherwise it is rejected.

Comparison of variances

To test the difference between the sample variance s^2 and a population variance σ^2, the test value

$$\chi_0^2 = (n - 1)s^2 / \sigma^2 \tag{C.28}$$

is first found. If $s^2 \leq \sigma^2$, then the test value χ_0^2 is compared with the critical value χ_{p1}^2 (Table 2 in ISO 12491[59]) corresponding to the $v = n - 1$ degrees of freedom and to the significance level $\alpha = p$. If $\chi_0^2 \geq \chi_{p1}^2$, then the hypothesis that the sample is taken from the population with the variance σ^2 is not rejected; otherwise it is rejected.

If $s^2 \geq \sigma^2$, then the test value χ_0^2 is compared with the critical value χ_{p2}^2 (Table 2 in ISO 12491[59]) corresponding to $v = n - 1$ degrees of freedom and to the significance level $\alpha = 1 - p_2$. When $\chi_0^2 \leq \chi_{p2}^2$, the hypothesis that the sample is taken from the population with the variance σ^2 is not rejected; otherwise it is rejected.

For two samples of sizes n_1 and n_2 the difference of the sample variances s_1^2 and s_2^2 (the subscripts are chosen such that $s_1^2 \leq s_2^2$) may be tested by comparing the test value

$$F_0 = s_1^2 / s_2^2 \tag{C.29}$$

with the critical value F_p, which is the fractile of the F-distribution given in Table 4 of ISO 12491[59] (see also ISO 2854[52]) for $v_1 = n_1 - 1$ and $v_2 = n_2 - 1$ degrees of freedom and for the significance level $\alpha = 1 - p$ (a small value close to 0). If $F_0 \leq F_p$ then the hypothesis that both samples are taken from populations with the same (though unknown) variance σ^2 is not rejected; otherwise it is rejected.

C.5. Estimation of fractiles

Introduction

From the probabilistic point of view, the characteristic value and the design value of a material strength can be defined as specific fractiles of the appropriate probability distribution. The fractile x_p is the value of a variable X satisfying

$$P(X \leq x_p) = p \tag{C.30}$$

where p denotes specified probability (see also expression (C.11)).

For the characteristic strength, the probability $p = 0.05$ is often assumed. However, for the design strength lower probabilities, say $p \approx 0.001$, are used. On the other hand the design value of non-dominating variables may correspond to greater probabilities, say $p \approx 0.10$.

Applied statistical techniques should be chosen cautiously, particularly when the design strength corresponding to a small probability is assessed. For example, when one is assessing the strength of concrete, only a very limited number of observations are usually available. Moreover, relatively high variability (coefficient of variation around 0.15) and, usually, a positive skewness can be expected. All these factors can produce a relatively high degree of statistical error.

Various statistical techniques for the assessment of characteristic and design strengths are available, corresponding to different assumptions concerning the type of distribution and the nature of available data. Commonly used methods assuming a normal distribution are well described in the new ISO document[59] and in a number of other publications. Generalised statistical techniques assuming a three-parameter distribution, where the coefficient of skewness is considered as an independent parameter, are available only in recent publications[39-41] for log-normal distribution. These documents are taken into account in the following brief review of the most important statistical methods.

Method of order

The most general method, which does not make any assumption about the type of distribution, is based on the statistics of order. A sample $x_1, x_2, \ldots, x_n$ is transformed into the ordered sample $x_1' \leq x_2' \leq \cdots \leq x_n'$, and the estimate $x_{p,\text{order}}$ of the unknown fractile x_p is

$$x_{p,\text{order}} = x_{k+1}' \tag{C.31}$$

where the index k follows from the inequality

$$k \leq np < k+1 \tag{C.32}$$

Obviously, this method needs a large number of observations if it is to be reliable.

Coverage method

The following classical technique (the so-called coverage method) is very frequently used.

The key idea is that the confidence level γ (often assumed to be 0.75, 0.90 or 0.95) for the lower p-fractile estimate $x_{p,\text{order}}$ is determined in such a way that

$$P(x_p, \text{cover} < x_p) = \gamma \tag{C.33}$$

If the population standard deviation σ is known, the lower p-fractile estimate $x_{p,\text{cover}}$ is given in terms of the sample mean m as

$$x_{p,\text{cover}} = m - \kappa_p \sigma \qquad (C.34)$$

If σ is unknown then the sample standard deviation s also needs to be used

$$x_{p,\text{cover}} = m - k_p s \qquad (C.35)$$

The coefficients κ_p and k_p depend on the type of distribution, on the probability p corresponding to the desired fractile x_p, on the sample size n, and on the confidence level γ. Explicit knowledge of γ, so that the estimate $x_{p,\text{cover}}$ will be on the safe side of the actual value x_p, is the most important advantage of the method. To take account of statistical uncertainty, the value for γ of 0·75 is recommended.[10] However, when an unusual reliability consideration is required, a higher confidence level 0·95 seems to be appropriate. Only the normal distribution is considered without taking into account any possible asymmetry of the population distribution.[10,59]

Prediction method

Another estimation method, developed assuming a normal distribution, is the prediction method. Here the lower p-fractile x_p is assessed by the prediction limit $x_{p,\text{pred}}$, determined in such a way that an additional value x_{n+1} randomly taken from the population would be expected to occur below $x_{p,\text{pred}}$ with the probability p, thus

$$P(x_{n+1} < x_{p,\text{pred}}) = p \qquad (C.36)$$

It can be shown that the prediction limit $x_{p,\text{pred}}$, defined by expression (C.36), does approach the unknown fractile x_p with increasing n, and from this point of view $x_{p,\text{pred}}$ can be considered as an approximation to x_p. If the standard deviation of the population σ is known, then $x_{p,\text{pred}}$ can be calculated from

$$x_{p,\text{cover}} = m - u_p \sigma (1/n + 1)^{1/2} \qquad (C.37)$$

where u_p is the p-fractile of the standardised normal distribution. If σ is unknown, then

$$x_{p,\text{cover}} = m - t_p s (1/n + 1)^{1/2} \qquad (C.38)$$

where t_p is the p-fractile of the known Student's t-distribution with $n - 1$ degrees of freedom.

Bayesian method

When previous observations of a continuous production are available, an alternative technique is provided by the Bayesian approach.[10,59] Let m be the sample mean and s the sample standard deviation, as determined from a sample of size n. In addition, from previous observations, assume that the sample mean m' and sample standard deviation s' are also known for a sample for which the individual results and the size n' are unknown. Both samples are assumed to be taken from the same population having theoretical mean μ and standard deviation σ. Hence the two samples can be considered jointly.

Parameters for the combination of the two samples are

$$n'' = n + n'$$
$$v'' = v + v' - 1 \text{ when } n' \geq 1, \ v'' = v + v' \text{ when } n' = 0$$
$$m'' = (mn + m'n')/n''$$
$$s''^2 = (vs^2 + v's'^2 + nm^2 + n'm'^2 - n''m''^2)/v'' \tag{C.39}$$

The unknown values n' and v' may be estimated using formulae for the coefficients of variation $V(m')$ and $V(s')$:

$$n' = \{\sigma/[\mu V(m')]\}^2, \ v' = 0{\cdot}5/[V(s')]^2 \tag{C.40}$$

n' and v' may each be chosen independently (here it does not hold that $v' = n' - 1$) with regard to previous experience concerning the degree of uncertainty in estimating the mean μ and standard deviation σ.

The Bayesian limit $x_{p,\text{Bayes}}$, considered as an assessment of x_p, is given by an expression similar to expression (C.38) used by the prediction method when σ is unknown

$$x_{p,\text{Bayes}} = m'' - t''_p s''(1/n'' + 1)^{1/2} \tag{C.41}$$

where t''_p is the p-fractile of Student's t-distribution for v'' degrees of freedom. Furthermore, if the standard deviation σ is known and sample data are used to determine mean values only, then $v = \infty$, s'' should be replaced by σ and instead of expression (C.41) the following formula, similar to expression (C.37) used by the prediction method when σ is known, should be used

$$x_{p,\text{Bayes}} = m'' - u_p s''(1/n'' + 1)^{1/2} \tag{C.42}$$

When one is applying the Bayesian technique for determining, for example, strength, advantage may be taken of the fact that long-term variability of strength is usually stable. Thus uncertainty in determining σ is relatively small, the value $V(s')$ is also small and v' given by expression (C.40) and v'' given by expression (C.39) are high. This may lead to a favourable decrease of the resulting value t''_p and to a favourable increase of the estimate of the lower fractile x_p (see expression (C.41)). On the other hand, uncertainty in determining μ and $V(m')$ is usually high and so any previous information might not significantly affect the value of n'' and m''.

If no prior information is available, then $n' = v' = 0$ and the characteristics m'', n'', s'', v'' equal the sample characteristics m, n, s, v. Expressions (C.41) and (C.42) now reduce to the previous expressions (C.37) and (C.38) respectively. In this special case the Bayesian approach leads to the same result as for the prediction method, and expressions (C.37) and (C.38) are to be used. It should be noted that this special case of Bayesian techniques with no prior information is considered in *Annex D* and in ISO documents.[10,59]

Comparison of coverage and prediction methods

When one is estimating the characteristic and design values, the coverage and prediction method are most frequently used. These methods are compared here by assuming a normal distribution for the population.

Table C.2 gives the coefficients κ_p and $u_p(1/n + 1)^{1/2}$ used in expressions (C.34) and (C.37) for selected values of n and γ. It follows from Table C.2 that differences between the two coefficients depend on the number of observations n as well as on the confidence level γ. For $\gamma = 0{\cdot}95$ and for small n the coefficient κ_p of the coverage method is almost 40% higher than the coefficient $u_p(1/n+1)^{1/2}$ of the prediction method. If $\gamma = 0.75$ is accepted,[10,59] the differences are less than 10%. Generally, however, the prediction method would obviously lead to higher (less safe) characteristic values than the classical coverage method for a confidence level $\gamma \geq 0.75$.

Table C.2. Coefficients κ_p and $u_p(1/n+1)^{1/2}$ for $p = 0.05$ and known σ

Coefficients		Number of observations n								
		3	4	5	6	8	10	20	30	∞
κ_p	$\gamma = 0.75$	2·03	1·98	1·95	1·92	1·88	1·86	1·79	1·77	1·64
	$\gamma = 0.90$	2·39	2·29	2·22	2·17	2·10	2·05	1·93	1·88	1·64
	$\gamma = 0.95$	2·60	2·47	2·38	2·32	2·23	2·17	2·01	1·95	1·64
$u_p(1/n+1)^{1/2}$		1·89	1·83	1·80	1·77	1·74	1·72	1·68	1·67	1·64

If the standard deviation σ is unknown, then expressions (C.35) and (C.38) need to be compared.

Table C.3 gives the appropriate coefficients k_p and $t_p(1/n+1)^{1/2}$ for the same number of observations n and confidence levels γ as in Table C.2. Obviously, differences between the coefficients corresponding to different confidence levels γ are much more significant than in the previous case of known σ. For $\gamma = 0.95$ and small n the coefficient k_p used by the coverage method is almost 100% greater than the coefficient $t_p(1/n+1)^{1/2}$ used by the prediction method. For $\gamma = 0.75$ the two coefficients are nearly the same. The coefficient k_p is, however, always slightly greater than $t_p(1/n+1)^{1/2}$ except for $n = 3$. As in the previous case of known σ, the prediction method would generally lead to greater (less safe) characteristic strengths than the classical coverage method. This difference increases with increasing confidence level.

Table C.3. Coefficients k_p and $t_p(1/n+1)^{1/2}$ for $p = 0.05$ and unknown σ

Coefficients		Number of observations n								
		3	4	5	6	8	10	20	30	∞
k_p	$\gamma = 0.75$	3·15	2·68	2·46	2·34	2·19	2·10	1·93	1·87	1·64
	$\gamma = 0.90$	5·31	3·96	3·40	3·09	2·75	2·57	2·21	2·08	1·64
	$\gamma = 0.95$	7·66	5·14	4·20	3·71	3·19	2·91	2·40	2·22	1·64
$t_p(1/n+1)^{1/2}$		3·37	2·63	2·33	2·18	2·00	1·92	1·76	1·73	1·64

Use of three-parameter log-normal distribution

Actual asymmetry of the population distribution may also have a significant effect on fractile estimation, particularly when small samples are taken from a population with high variability.

Assuming a general three-parameter log-normal distribution with an independent coefficient of skewness α, operational statistical techniques can often be effectively used when assessing material properties.

The effect of population asymmetry on the 0·05-fractile estimate is given in Tables C.4 and C.5 for two confidence levels and for three coefficients of skewness $\alpha = -1$, 0 and +1. Only the coverage method for the situation where σ is unknown, described above by expression (C.35), is considered. Table C.4 gives the coefficient k_p for a selection of numbers of observations n and for a confidence level $\gamma = 0.75$. Table C.5 gives the coefficient k_p for the same numbers of observations n as in Table C.4, but for the confidence level $\gamma = 0.95$.

Table C.4. Coefficients k_p for $p = 0.05$, $\gamma = 0.75$ and unknown σ

Coefficients of skewness	Number of observations n								
	3	4	5	6	8	10	20	30	∞
$\alpha = -1.00$	4·31	3·58	3·22	3·00	2·76	2·63	2·33	2·23	1·85
$\alpha = 0.00$	3·15	2·68	2·46	2·34	2·19	2·10	1·93	1·87	1·64
$\alpha = 1.00$	2·46	2·12	1·95	1·86	1·75	1·68	1·56	1·51	1·34

Table C.5. Coefficients k_p and for $p = 0.05$, $\gamma = 0.95$ and unknown σ

Coefficients of skewness	Number of observations n								
	3	4	5	6	8	10	20	30	∞
$\alpha = -1.00$	10·09	7·00	5·83	5·03	4·32	3·73	3·05	2·79	1·85
$\alpha = 0.00$	7·66	5·14	4·20	3·71	3·19	2·91	2·40	2·22	1·64
$\alpha = 1.00$	5·88	3·91	3·18	2·82	2·44	2·25	1·88	1·77	1·34

Using data from Table C.5, the coefficient k_p is shown in Fig. C.1 as a function of n for the three coefficients of skewness α. Comparing data given in Tables C.4 and C.5, it should follow that the effect of asymmetry on the estimate $x_{p,\text{cover}}$ increases considerably with increasing confidence level γ. Generally the effect decreases with increasing n, but it never vanishes even for $n \to \infty$. Detailed analysis[41] shows that when one is assessing the characteristic strength of a material property corresponding to the 0·05-fractile, actual asymmetry of probability distribution should be considered whenever the coefficient of skewness is greater (in absolute value) than 0·5.

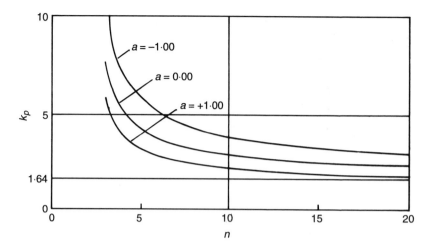

Fig. C.1.   Coefficients k_p for $p = 0.05$, $\gamma = 0.95$ and unknown σ

The differences between estimates obtained by using a general log-normal distribution with a given coefficient of skewness $\alpha \neq 0$, and the corresponding estimates based on a normal distribution with $\alpha = 0$, increase with decreasing probability p. Therefore, the design value of concrete strength of a material property corresponding to a very small probability p (say 0·001) should be determined directly from test data only in cases where

(i)   a sufficient number of observations
(ii) convincing evidence concerning the appropriate probabilistic model (including information on asymmetry)

are available.

When such evidence is not available, the design value should preferably be determined by assessing a characteristic value, to be divided by a partial factor and possibly by an explicit conversion factor.

Worked example C.1

A sample of concrete strength measurements of size $n = 6$, having $m = 23.9$ MPa and $s = 4.2$ MPa is to be used to assess the characteristic value of the strength $f_{ck} = x_p$ where $p = 0.05$.

Using the coverage method, it follows from expression (C.6) and Table C.3 that for the confidence level $\gamma = 0.75$

$$f_{ck,cover} = 23.9 - 2.34 \times 4.2 = 14.1 \text{ MPa} \tag{C.43}$$

and, for the confidence level $\gamma = 0.95$

$$f_{ck,cover} = 23.9 - 3.71 \times 4.2 = 8.3 \text{ MPa} \tag{C.44}$$

If the prediction method is used, it follows from expression (C.38) and Table C.3 that

$$f_{ck,pred} = 23.9 - 2.18 \times 4.2 = 14.7 \text{ MPa} \tag{C.45}$$

Thus, using the prediction method (which is identical with the Bayesian technique when no prior information is available[10,59]), the characteristic strength is almost 80% higher than the strength determined by the coverage method with confidence level 0.95.

When information from previous experience is available, the Bayesian approach can be used. Assume the following prior information

$$s' = 4.4 \text{ MPa}, \ V(s') = 0.28 \qquad (C.46)$$

It follows from expression (C.40) that

$$n' = [4.4/(25.1 \times 0.50)]^2 < 1, v' = 0.5/0.28^2 \approx 6 \qquad (C.47)$$

So the following characteristics are then used: $n' = 0$ and $v' = 6$. Taking account of the fact that $v = n - 1 = 5$, expressions (C.39) yield

$$n'' = 6, v'' = 11, m'' = 23.9 \text{ MPa}, s'' = 4.3 \text{ MPa} \qquad (C.48)$$

and, finally, it follows from expression (C.41) that

$$f_{ck,Bayes} = 23.9 - 1.8 \times 4.3 \times (1 + 1/6)^{1/2} = 15.5 \text{ MPa} \qquad (C.49)$$

where the value t_p'' is taken from tables of the Student's t-distribution. The resulting characteristic strength is therefore greater (by 0.8 MPa) than the value obtained by the prediction method. Also, other available information on the application of the Bayesian approach clearly indicates (see Annex D of ISO 2394[10]) that when previous experience is available this technique can be effectively used. This is particularly so in the case of a relatively high variability of concrete quality for the assessment of existing structures.

For some materials such as normal-strength concrete, a positive asymmetry of the probability distribution (with a coefficient of skewness of up to 1) is often observed. The following example of the effect of such skewness assumes that the sample of $n = 6$ concrete strength measurements, analysed above, is taken from a population with a log-normal distribution having a coefficient of skewness equal to 1. Using the classical coverage method for the confidence level $\gamma = 0.75$, expression (C.35) and Table C.4 yield

$$f_{ck,cover} = 23.9 - 1.86 \times 4.2 = 16.1 \text{ MPa} \qquad (C.50)$$

For the confidence limit $\gamma = 0.95$, expression (C.35) and Table C.5 yield

$$f_{ck,cover} = 23.9 - 2.82 \times 4.2 = 12.1 \text{ MPa} \qquad (C.51)$$

These values are greater by 14% and 46% respectively than in the previous cases where asymmetry was disregarded; therefore, by taking positive asymmetry into account, more favourable estimates are obtained. Similar effects of asymmetry on characteristic and design strengths are to be expected if the prediction and Bayesian methods are used. It should be noted that possible negative asymmetry, which may occur in case of high-strength materials, would correspondingly cause an unfavourable effect on the resulting values.

APPENDIX D

National Standard Organisations

It is stated in *clause 24* to the *Foreword* of *Eurocode 1, Part 1, Basis of Design:* 'It is intended that, during the ENV period, this prestandard is used for design purposes, in conjunction with the particular National Application Document valid in the country where the designed structures are to be located'.

Users of Eurocode 1, Basis of Design may obtain information with regard to the availability of a particular National Application Document from the relevent National Standard Authority listed below.

CEN National Members

Austria   Österreichisches Normungsinstitut (ON)
Postfach 130
Heinestraße 38
A-1021 Wien

Telephone: + 43 1 213 00
    Fax: + 43 1 213 00 650

Belgium   Institut Belge de Normalisation/
    Belgisch Instituut voor Normalisatie (IBN/BIN)
Avenue de la Brabançonne 29/
Brabançonnelaan 29
B-1040 Bruxelles/Brussel

Telephone: + 32 2 738 00 90
    Fax: + 32 2 733 42 64

Denmark   Dansk Standard (DS)
Baunegaardsvej 73
DK-2900 Hellerup

Telephone: + 45 39 77 01 01
    Fax: + 45 39 77 02 02

Finland	Suomen Standardisoimisliitto r.y. (SFS) PO Box 116 FIN-00241 Helsinki Finland Telephone: + 358 0 149 93 31 Fax: + 358 0 146 49 25
France	Association Française de Normalisation (AFNOR) Tour Europe F-92049 Paris la Défense Telephone: + 33 1 42 91 55 55 Fax: + 33 1 42 91 56 56
Germany	Deutsches Institut für Normung e.V. (DIN) D-10772 Berlin Telephone: + 49 30 26 01 0 Fax: + 49 30 26 01 12 31
Greece	Hellenic Organization for Standardization (ELOT) 313, Acharnon Street GR-11145 Athens Telephone: + 30 1 228 00 01 Fax: + 30 1 202 07 76
Iceland	Icelandic Council for Standardization (STRI) Keldnaholt IS-112 Reykjavik Telephone: + 354 587 70 02 Fax: + 354 587 74 09
Ireland	National Standards Authority of Ireland (NSAI) Glasnevin IRL-Dublin 9 Telephone: + 353 1 807 38 00 Fax: + 353 1 807 38 38
Italy	Ente Nazionale Italiano di Unificazione (UNI) Via Battistotti Sassi, 11b I-20133 Milano MI Telephone: + 39 2 70 02 41 Fax: + 39 2 70 10 61 06
Luxembourg	Inspection du Travail et des Mines (ITM) Boite Postal 27 26, rue Zithe L-2010 Luxembourg Telephone: + 352 478 61 50 Fax: + 352 49 14 47

Netherlands Nederlands Normalisatie Instituut (NNI)
Postbus 5059
Kalfjeslaan 2
NL-2600 GB Delft

Telephone: + 31 15 269 03 90
Fax: + 31 15 269 01 90

Norway Norges Standardiseringsforbund (NSF)
PO Box 353 Skoyen
N-0212 Oslo

Telephone: + 47 22 04 92 00
Fax: + 47 22 04 92 11

Portugal Instituto Portugues da Qualidade (IPQ)
Rua C, Av. dos Tres Vales
P-2825 Monte da Caparica

Telephone: + 351 1 294 81 00
Fax: + 351 1 294 81 01
+ 351 1 294 82 22

Spain Asociación Española de Normalización y Certificación
(AENOR)
Calle Fernández de la Hoz, 52
E-28010 Madrid

Telephone: + 34 1 432 60 00
Fax: + 34 1 310 49 76

Sweden Standardiseringen i Sverige (SIS)
Box 6455
S-113 82 Stockholm

Telephone: + 46 8 610 30 00
Fax: + 46 8 30 77 57

Switzerland Schweizerische Normen Vereinigung (SNV)
Mühlebachstraße 54
CH-8008 Zurich

Telephone: + 41 1 254 54 54
Fax: + 41 1 254 54 74

United Kingdom British Standards Institution (BSI)
389 Chiswick High Road
GB-London W4 4AL

Telephone: + 44 181 996 90 00
Fax: + 44 181 996 74 00

CEN Affiliates for CEN/TC 250 (August 1996)

Bulgaria
Committee for Standardization and Metrology (CSM)
21, rue du 6 Septembre
BG-1000 Sofia

Telephone: + 359 2 88 58 98
Fax: + 359 2 80 14 02

Croatia
State Office for Standardization and Metrology (DZNM)
Ulica grada Vukovara 78
HR-41000 Zagreb

Telephone: + 385 1 539 934
Fax: + 385 1 536 598

Czech Republic
Czech Office for Standards, Metrology and Testing (COMST)
Biskupský Dvůr 5
CS-113 47 Praha 1, Czech Republic.

Telephone: + 42 2 232 44 30
Fax: + 42 2 232 43 73

Poland
Polish Committee for Standardization (PKN)
ul. Elektoralna 2
PL-00-139 Warszawa

Telephone: + 48 2 620 54 34
Fax: + 48 2 620 07 41

Slovenia
Standards and Metrology Institute (SMIS)
Kotnikova 6
SI-61000 Ljubljana
Republic of Slovenia

Telephone: + 386 61 13 12 322
Fax: + 386 61 31 48 82

References

1. CEN TC/250. Project Team Eurocode 1.1. *Background Documentation Eurocode 1: Part 1: Basis of Design.* ECCS, March 1996.
2. Beeby A.W. and Narayanan R.S. *Designer's handbook to Eurocode 2: Part 1.1: Design of concrete structures*, Thomas Telford, 1993.
3. CEN TC/250. *Statement of Intent of CEN/TC 250 Structural Eurocodes.* Document CEN/TC 250 N 211, November 1994.
4. CEN TC/250 Interim AD-HOC Group. *Basis of Design — Transposition of ENV 1991-1 to EN 1991-1.* Document CEN/TC 250 N 248, September 1995.
5. The Institution of Structural Engineers. *Appraisal of existing structures.* ISE, 1980.
6. Building Research Establishment. *Structural appraisal of existing buildings for change of use.* BRE Digest 366, Garston Watford. October 1991.
7. Société suisse des ingénieurs et des architectes. *Evaluation de la sécurité structurale des ouvrages existants.* SIA Directive 462, Zürich, August 1994.
8. Czech Office for Standards, Metrology and Testing. *Design and assessment of building structures subjected to reconstruction (Navrhování a posuzování stavebních konstrukcí při přestavbách).* ÚNM Praha, 1987, ČSN 730038-1988.
9. International Standards Organisation. *General principles on reliability for structures — List of equivalent terms, Trilingual edition.* ISO, Geneva, 1987, ISO 8930.
10. International Standards Organisation. *General principles on reliability for structures.* ISO, Geneva, 1988, ISO 2394.
11. International Standards Organisation. *Bases for design of structures — Notations — General symbols.* ISO, Geneva, 1987, ISO 3898.
12. Department of the Environment and The Welsh Office. *Approved Document A (Structure) to the Building Regulations 1991.* HMSO, London, 1985.
13. British Standards Institution. *Code of practice for use of masonry: Part 1: Structural use of unreinforced masonry.* BSI, London, 1992, BS 5628 : Part 1.
14. British Standards Institution. *Structural use of steel work in building: Part 1: Code of practice for design in simple and continuous construction: hot rolled sections.* BSI, London, 1990, BS 5950 : Part 1.
15. British Standards Institution. *Structural use of concrete: Part 1: Code of practice for design and construction.* BSI, London, 1985, BS 8110: Part 1.
16. British Standards Institution. *Structural use of concrete: Part 2: Code of practice for special circumstances.* BSI, London, 1985, BS 8110: Part 2.
17. British Standards Institution. *Structural use of timber: Part 2: Code of practice for permissible stress design, materials and workmanship.* BSI, London, 1991, BS 5268 : Part 2.
18. British Standards Institution. *Eurocode 1: Part 1: Basis of design (includes UK NAD).* BSI London, October 1996, prENV 1991-1.
19. International Standards Organisation. *Quality management and quality assurance standards — Part 1: Guidelines for selection and use.* ISO, Geneva, 1994, ISO 9000-1.

20. International Standards Organisation. *Quality management and quality assurance standards — Part 2: Generic guidelines for the application of ISO 9001, ISO 9002 and ISO 9003.* ISO, Geneva, 1993, ISO 9000-2.

21. International Standards Organisation. *Quality management and quality assurance standards — Part 3: Guidelines for the application of ISO 9001 to the development, supply and maintenance of software.* ISO, Geneva, 1991, ISO 9000-3.

22. International Standards Organisation. *Quality management and quality assurance standards — Part 4: Guide to dependability programme management.* ISO, Geneva, 1993, ISO 9000-4.

23. International Standards Organisation. *Quality systems — Model for quality assurance in design, development, production, installation and servicing.* ISO, Geneva, 1994, ISO 9001.

24. International Standards Organisation. *Quality systems — Model for quality assurance in production, installation and servicing.* ISO, Geneva, 1994, ISO 9002.

25. International Standards Organisation. *Quality systems — Model for quality assurance in final inspection and test.* ISO, Geneva, 1994, ISO 9003.

26. International Standards Organisation. *Quality management and quality system elements — Part 1: Guidelines.* ISO, Geneva, 1994, ISO 9004-1.

27. International Standards Organisation. *Quality management and quality system elements — Part 2: Guidelines for services.* ISO, Geneva, 1991, ISO 9004-2.

28. International Standards Organisation. Quality management and quality system elements — Part 3: Guidelines for processed materials. *ISO, Geneva, 1993, ISO 9004-3.*

29. International Standards Organisation. *Quality management and quality system elements — Part 4: Guidelines for quality improvement.* ISO, Geneva, 1993, ISO 9004-4.

30. Holický M. *Fuzzy Concept of Serviceability Limit States.* Symposium/Workshop on Serviceability of Buildings, NRC Canada, Ottawa, Canada, 1988, p. 19.

31. Holický M. *Fuzzy Criteria of Serviceability Limit States.* Acta Polytechnica. Vol. 11 (I,4), 1990, p. 237–325.

32. Holický M. and Östlund L. *Probabilistic Design Concept.* CIB International Colloquium on Structural Serviceability, Goteborg, Sweden, June 1993, p. 91.

33. Deak G. and Holický L. *Serviceability Requirements.* CIB International Colloquium on Structural Serviceability, Goteborg, Sweden, June 1993, p. 25.

34. Holický M. *Fuzzy Optimisation of Structural Reliability.* ICOSSAR 93, Innsbruck, August 1993, p. 1379.

35. International Standards Organisation. *Bases for design of structures — Serviceability of structures against vibrations.* ISO, Geneva, 1992, ISO 10137.

36. International Standards Organisation. *Evaluation of human exposure to whole-body vibration Part 1: General Requirements.* ISO, Geneva, 1985, ISO 2631:1.

37. International Standards Organisation. *Evaluation of human exposure to whole-body vibration — Part 2: Continuous and shock-inducted vibrations in buildings (1-80 Hz).* ISO, Geneva, 1989, ISO 2631: 2.

38. International Standards Organisation. *Evaluation of human exposure to whole-body vibration Part 3: Evaluation of exposure to whole-body z-axis vertical vibration in the frequency range 0,1to 0,63 Hz.* ISO, Geneva, 1985, ISO 2631:3.

39. Holický M. and Vorlíček M. *Fractile Estimation and Sampling Inspection in Building.* Acta Polytechnica, Vol. 32, No. 1/1992, p. 87.

40. Holický M. and Vorlíček M. *Distribution Asymmetry in Structural Reliability.* Acta Polytechnica, Vol. 35, No. 3/ 1995, p. 75.

41. Holický M. and Vorlíček M. General Lognormal *Distribution in Statistical Quality Control.* Application of Statistics and Probability, ICASP 7, Paris, 10 – 13 July 1995. A.A Balkema, Rotterdam 1995, p. 719.

42. Vorlíček M. and Holický M. *Analysis of Dimensional Accuracy for Building Structures.* Elsevier, Netherlands, 1989, p. 260.

43. Holický M. *Time Dependent Dimensional Deviations in Accuracy Analysis.* CIB Working Papers, Publication 112, Rotterdam, Netherlands, 1989, p. 28.

44. Holický M. and Holická N. *Serviceability and Tolerances.* In: CIB International Colloquium on Structure Serviceability, Goteborg, Sweden, June 1993, p. 33.

45. British Standards Institution. *Guide to accuracy in building*. BSI, London, 1990, BS 5606.
46. Breitschaft G., Östlund L. and Kersken-Bradley M. *The Structural Eurocodes — Conceptual Approach*. IABSE Conference: Structural Eurocodes, p.10, Davos 1992.
47. Holický M. and Vorlíček M. *Statistical Procedures for Design Assisted by Testing*. In: IABSE Colloquium Basis of Design and Actions on Structures, Delft 1996, p. 317.
48. Johnson R.P. and Anderson D. *Designer's handbook to Eurocode 4: Part 1.1: Design of composite steel and concrete structures*, Thomas Telford, 1993.
49. European Commission. *Construction Products Directive (Directive 89/106/EEC)*. Official Journal of the European Communities (No. L/40 of 11. 2. 1989, pp.12–26).
50. European Commission. *General introduction to the 6 Interpretative Documents (IDs) and ID No 1: Mechanical resistance and stability*. Official Journal of the European Communities (No. C 62/1, 28.2.94, pp.1–22).
51. International Standards Organisation. *Statistical interpretation of test results. Estimation of the mean. Confidence interval*. ISO, Geneva, ISO 2602.
52. International Standards Organisation. *Statistical interpretation of data. Techniques of estimation and tests relating to means and variances*. ISO, Geneva, ISO 2854.
53. International Standards Organisation. *Statistical interpretation of data. Determination of a statistical tolerance interval*. ISO, Geneva, ISO 3207.
54. International Standards Organisation. *Statistical interpretation of data. Comparison of two means in the case of paired observations*. ISO, Geneva, ISO 3301.
55. International Standards Organisation. *Statistics. Vocabulary and symbols*. ISO, Geneva, ISO/DIS 3534 : Part 1.
56. International Standards Organisation. *Statistics. Vocabulary and symbols*. ISO, Geneva, ISO/DIS 3534 : Part 2.
57. International Standards Organisation. *Sampling procedures and charts for inspection by variables for percent non-conforming*. ISO, Geneva, ISO 3951.
58. International Standards Organisation. *Normality tests*. ISO, Geneva, ISO/DIS 5479.
59. International Standards Organisation. *Statistical methods for quality control of building materials and components*. ISO 12491.

Index

Variable action(s), 14, 35, 36, 39, 70
Variance, 98, 99, 103, 104
Verification(s), 68, 74, 81
Vertical ties, 17
Vibration, 35
  criteria, 33

Wind, 21
Wind action(s) (load(s)), 22, 35, 38
Wind oscillation(s), 10, 34
Workmanship, 11

Yield stress, 44